Dictionary of Military Architecture

Fortification, Fieldworks and Siegecraft

2nd Edition

By Stephen (Sven) Wyley.

MELBOURNE

Front Cover. The Turrets in ink, pencil and watercolour (Observation tower, Fort Nepean, Victoria, Australia) by Katie Roberts (2017).

"*Attention. Permissions Coordinator,*" at the address below.
Wyley. C/- Intertype,
Unit 45, 125 Highbury Road,
BURWOOD, VIC, 3125, Australia.
www.intertype.com.au

Ordering Information.
Quantity sales. Special discounts are available on quantity purchases by corporations, associations, and others. For details, contact the "*Special Sales Department*" at the address above.

The Dictionary of Military Architecture, Fortification, Fieldworks and Siegecraft. —2nd edition.
ISBN 978-1-7642798-7-1

Acknowledgements.

I would like to thank the following people for their support, encouragement, impetus, and contributions to this work; Steven Baker, Martin Bonnici, Leigh Campbell, Elizabeth Cowling, Mike Dunajew, Alex Foster, Leah Godfrey, Rashmi Gune, Michelle Holian, Jim Kaufmann, Sean Keenan, Fathma Nachair, Dave Newman, John Slone, Adam Turner, Dom Williams, Jake Whistance.

Thanks to the artists who contributed their works to help illustrate this book, including; Katie Roberts, Steven Lowe and Gary Bannister.

Thanks to the following for the photographs for the use of their work; Sarah and Michael Brown, Awbrie Culleton, Dr. Timothy Dawson, Damien Fegan, Adrian and Hannah Ferrer, Alan Greenway, Naomi Haddrick, Ruth Kunde, Wendy Lale, Jonathon Lyon, Jason Metcalfe, Kristen Pincott, Katie Roberts, Glenda and Wayne Robinson, Steve Sinclair, Aaron Southwell, and Kim and Graham Walker.

Special thanks to the following people for taking their red pen to my manuscript; Rey Croucher, Bret Kearney and Wayne Robinson.

Thanks to all the writers and researchers that have published their works on military architecture, castles and siegecraft, as they say '*on the shoulders of giants*'.

Stephen '*Sven*' Wyley
2025

Select internal illustrations by Katie Roberts,

(Katiebell_illustration). Dedicated to the hope that warfare becomes an artifact of the past; to peace and to all those who continue to suffer from war, especially women and children.

Table of Contents.

"The general must be the first in the toils and fatigues of the army. In the heat of summer he does not spread his parasol nor in the cold of winter don thick clothing. In dangerous places he must dismount and walk. He waits until the army's wells have been dug and only then drinks; until the army's food is cooked before he eats; until the army's fortifications have been completed, to shelter himself."

- Quote from Sun Tzu.
General, strategist, philosopher, and writer.
(771 – 256 BCE)

Preface.

After some more research and prompting by numerous readers I have succumbed to pressure to publish a 2nd edition of this book, with a large number of new entries, including more terms from Japan, China, India and Germany. Along with more terms there are more drawings and photographs, with the addition of two articles, one on the *'Designing and building your own European castle'*, and the other on *'linear fortifications'*. It is my hope this edition will go on to aid others in their writing and research.

There are a number of reasons why I was attracted to military architecture. In my past I had a hand in designing and building a number of forts for living history projects, where we would build them and then fight over them. So I wrote a number of articles on castles for newsletters and other publications. Then I came across a book, *'Classic Dictionary of Architecture'*, A concise glossary of terms used in Grecian, Roman, Italian, and Gothic architecture by John Henry Parker (*1846 CE*). And it was this book that inspired me to write my own dictionary but on fortifications.

I started this project in the late 1980's CE and put it on the internet for all to see in the 1990's CE, updating and adding items along the way, the last update to the '*Angelfire*' site was in 2003, later it was moved to '*Geocities*' then '*Google*' thanks to Wayne Robinson (which was last updated in 2005), thus redundant copies are still accessible there.

Military architecture has developed over thousands of years and has been used worldwide to defend and attack fortified borders, cities, ports, towns, villages, castles and forts. Most of the extant ancient works of the Egyptians, Hittites, Greeks and Romans origin, using materials as simple earth, turf, stone, brick, timber, iron and concrete. So this book covers the military architecture, fortifications, fieldworks and siegecraft of Europe, the Americas and Asia and everywhere in between.

The elements of military architecture have increased in sophistication over time with changes in modes and machines of attack and defence from escalade to mining, from trebuchet to cannon, siege engines and battering rams to tanks, ditch digging machines, missiles and drones.

Omissions from this text include details of: artillery (guns and cannons); and ordinance (bunker buster missiles to '*V*' bombs); from boats to ships; from airplanes to zeppelins; and tanks. I agree that '*military architecture evolves to respond to each innovation in ordinance*'[1] but ordinance and delivery systems are beyond the realms of this project.

As those involved in the 2nd winter of the Ukrainian/Russia war will attest, warfare is not glamorous or fun, we are all losers. However, we need to know the past to learn from it and attempt to make a better world.
In this dictionary I have attempted to provide a text which explains many of the terms associated with fortifications, field works and siege warfare, which were used between the ancient times to the present, covering Asia, Europe and the Americas. There are a great number of texts on the subject of castles, fortifications and fieldworks, let alone siege craft but few, if any, modern texts to my knowledge which specifically explain the terminology used to describe them.

With this book I have not sought to provide a preferred definition but to present how the terms have been used in the past, and how they are used today. The objective of this book is for helpful reference for military historians, historical re-enactors, war gamers, cosplayers, fiction writers, and for study or discussion concerning the development and the historical significance of military architecture, and therefore all words and meanings pertaining to the subject qualify for inclusion.

Not everyone is likely to agree completely with my selection of terms, likewise, there will be disagreements over some of the definitions, particularly where I had to be arbitrary in my selection between conflicting sources. Nevertheless, I accept responsibility for all errors and omissions, and would welcome notice of them.

Stephen '*Sven*' Wyley
2025

---

1 '*...military architecture evolves to respond to each innovation in ordinance*'. Wayne Robinson, personal correspondence 26/7/2024.

Notes on use.
The entries are in strict alphabetical order, ignoring all punctuation and word breaks. Reversed phases are listed only when the natural order might be hard to find.

Etymology.
In general, where the origin of the entries can be traced with some certainty, the language is indicated at the start of the entry. The statement of the origin is written as follows;
(Fr. basse, court),
1. 2. 3,
1. The language of origin,
2. The actual form of the word itself or its chief element in that language,
3. The literal meaning in that language or a brief indication of the sense involved.

The abbreviations of the most common languages are as follows:

(Ar.) Arabic.	(H.) Hindustani
(Ch.) Chinese	(J.) Japanese
(Du.) Dutch	(L.) Latin
(E.) English	(N.) Norse
(Fr.) French	(O.E) Old English
(Gaelic) Gaelic	(O.Fr) Old French
(G.) German	(Sc.) Scottish
(Gr.) Greek	(Sp.) Spanish
(Ir.) Irish	(Teut.) Teutonic
(It.) Italian	Note: For other languages the full name is used.

A.

A-barba. Guns sited to fire over the tops of a parapet.

Abatic, Abatis, Abattis. (O.Fr. *enbatre*, to thrust in). A field defence or rampart which was made of felled trees with the sharpened boughs facing outwards, towards the direction of attack.

See 'astverhau', 'baumstämmen', 'obstacle', 'ostrog', 'panzersperre aus aumstämmen', 'pfahlsperre', 'zasechnaya cherta'.

Abri d'intervalle. (Fr. *interval shelter*). A shelter on the *'Maginot Line'* for troops defending the area between the main works.

See 'Maginot line'.

Abschnittsgraben. (G. *section ditch*). Cross ditches or moats, separating the outer and inner wards or baileys.

Abwurfdach. (G. *drop roof*). An easily removable roof on top of the curtain walls, cavaliers and bastions of a bergfried, when the roof was removed catapults could be erected to defend the castle, dating around 1550 CE.

Acropolis. (Gr. *Akropolis*). Ancient Greek hilltop stronghold, using natural defences of a site and stone walls *(i.e. 'The Acropolis of Athens').*

Admiralty scaffolding. A type of anti-tank/boat obstacle, made from tubular steel, designed by the British and deployed on parts of the shores of England during the threat of invasion by Germany in 1940-1941 CE. Also known as *'obstacle Z.1'*, *'beach scaffolding'*, *'anti-tank scaffolding'*.

See 'beach obstacle', 'obstacle'.

Adulterine castle. The unlicensed castles (probably of the 'motte and bailey' design) were constructed without royal permission by the Barons supporting Matilda in England during the reign of King Stephen (1136-1152 CE), and demolished by King Henry II after the treaty of 1153 CE. An example was King Stephen's brother Henry, Bishop of Winchester who had to destroy the three castles he had built near the cathedral city.

Advanced posts. Entrenched works forward of the main line provided with means of safe withdrawal, but open to fire from the main line, used to provide flanking fire over dead spaces or to occupy territory that should be denied to the enemy.

See *'outwork', 'ravelin'.*

Advanced armour. Additional protection for a 19th century CE gun turret consisting of an outer ring of cast iron.

See 'armoured turret', 'turret'.

Advanced ditch. A ditch dug in front of the glacis sat opposite to the covered way.

Advanced glacis. ***See 'glacis'.***

Advanced post. A small work forward of the main line used to warn of and delay a surprise attack.

See 'advanced work', 'advanced line', 'antemural', 'antemuralia', 'arrow', 'barbican', 'counterguard', 'crownwork', 'detached work', 'forework', 'fléche', 'outwork', 'ravelin', 'fortalice', 'hornwork', 'horse shoe'. 'moon', 'multivallate hillfort', 'paté', 'redoubt', 'russian sap', 'vamure', 'yue cheng', 'zwinger'.

Advanced work. An outer defence of a fortification situated beyond the glacis but placed close enough to receive covering fire from the main works.

See *'outwork', 'ravelin'.*

Agger, Aggeri. (L. *mound*).

(1) A siege ramp. An example is the earthen ramp used by the Romans to attack Masada (Israel).

(2) The rampart formed by heaping the excavated soil up on the inner side of a ditch surrounding a Roman fortification.

See 'ditch'.

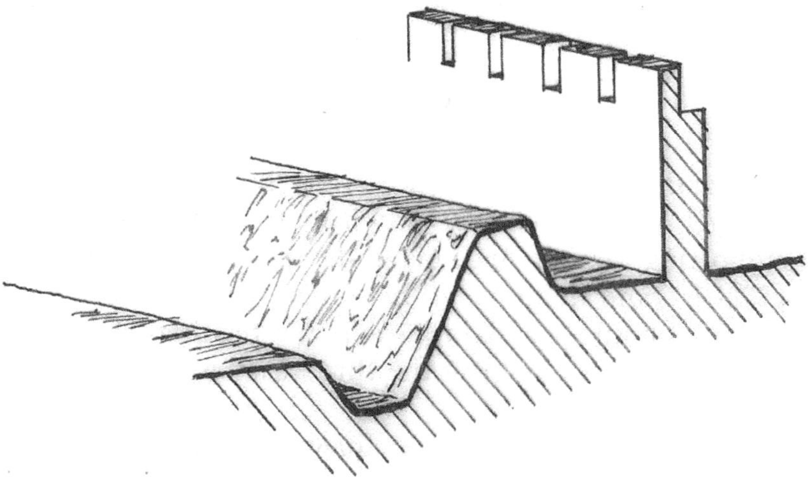

Aile. A wing or flank of a fortification.

Air hole. A hole to extract smoke from gun fire from a loop hole, embrasure or casemate.

Airina durga. (Sanskrit). A fort built on a saline soil or swamp of saline water and protected by thorny bushes.

See 'durga'.

Air raid shelter. A place or building used to find refuge during an air raid or bombardment, which can be purpose built or improvised.

See *'Anderson hut', 'bunker', 'flak towers (Germany)', 'flakturm', 'hochbunker', 'luftschutzraum', 'London deep-level shelters', 'Morrison shelter', 'Stanton air raid shelter', 'Winkel shelter'.*

Aiyokoya-no-sama. (J.). A loophole built in a castle walls or ramparts to provide crossfire on enemy soldiers from bends in the walls or ramparts, for loosing arrows and firing muskets. Located so missiles did not threaten defenders on the opposite walls or rampart.

See *'sama'.*

Akashjanani. (Sanskrit). A thin wall on a parapet of a fort with holes to observe the attackers.

Akijiro. (J. *empty castle*). An empty castle because the lord died or the garrison has retreated from the castle.

See *'jiro'.*

Alatorium. (L.). The wallwalk behind the battlements of a fortification, which enables the defenders to protect their position from the tops of the walls. Also known as an allure or parapet walk.

See 'allure'.

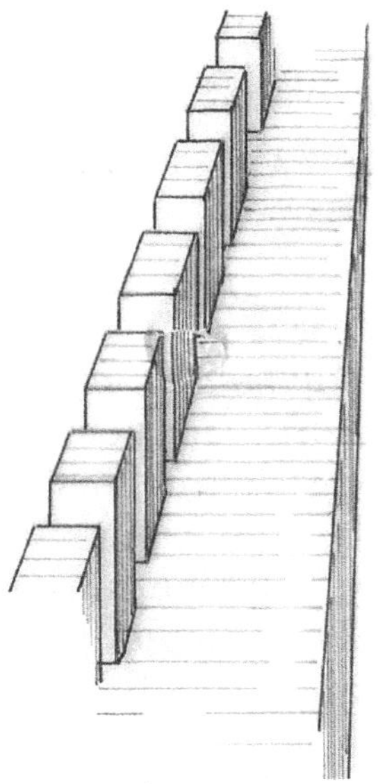

Albarrana. (Ar. *barranu,* exterior).

(1) A bridge connecting separate towers of Andalusian fortifications. An example is the Malmuerta Tower, in Córdoba, Spain.

(2) A tower connected to the curtain wall by a bridge.

An example is the Torre de Espantaperros in Badajoz, Spain.

See 'kong jun tai'.

Alcazaba. (Ar. *al-Qasr*). A stronghold built by the Saracens in Spain, 5th to the 8th century CE. The irregular walls of the enceinte followed the contours of a defensible hill, and the curtain walls were reinforced by square or polygonal towers. The towers were constructed chiefly of '*tapia*', a mixture of cement and pebbles poured between boards, thus the material was not suitable for round towers, producing square and polygonal shapes.

Alcasar, Alcazar, Alkazar. (Ar. *al-Qasr*) When a Muslim Alcazaba was rebuilt and enlarged in stone it became an alcazar. These were often used as a fortified palace of the regional military governor. An example is the Alcazar of Segovia, Spain.

Allan Williams Turret. A British WWII prefabricated metal armoured turret that sat atop a concrete or brick lined pit dug into the ground, provided with an armoured turret with a slot for firing a weapon and a top hatch for escape or providing anti-aircraft fire, mostly used on RAF airfields.

See 'FW3'.

Allure, Alur, Alure. (O.E. *alours*). A passage or gallery behind a parapet at the top of a wall of a fortification. Also known as a bailey walk, parapet walk, wall walk or wall walkway.

See 'alatorium'.

Al-Qasr. ***See 'alcasar'.***

Almenara. (*Ar. watchtower*). A small fort used as a watchtower on the Iberian Peninsula.

See 'atalaya'.

Alaska barrier. See *'bremer wall'.*

Alpine Wall. (It. *Vallo Alpino del Littono*). An Italian system of fortifications along Italy's north frontier dating from 1930 to 1942 CE. (1,851 km (1,150 mile)) long. This defensive line faced Austria, France, Switzerland and Yugoslavia. The Alpine Wall was defended by block houses armed with artillery, machine guns, mortars and flamethrowers. The combat blocks were connected to support areas in the rear consisting of; ammunition stores, barracks, command centres and utility rooms.

See *'Rupnik Line'.*

Alternating staircase. A staircase which was used for the movement between different floors of a tower or keep. The staircase reversed its turn from right-to-left to left-to-right from floor to floor. This design was used to break the flow of the enemy's advance once they had gained entry to the keep.

American West forts. ***See 'cavalry forts'.***

Ammunition store. A storage area which allows both the swift supply of ammunition to a gun whilst being as safe as possible, in later fixed positions the ammunition was supplied to the gun with power drive hoist.

See 'arsenal', 'enshõgura', 'expense magazine', 'field magazine', 'filling room', 'magazine', 'polverista', 'principal magazine', 'pulverturm'.

Anastasian wall. A linear defence consisting of a ditch, outer works, gates and forts, raised under the emperor Anastasius I in 507 CE to stop the attacks of the Bulgarians, and continued in use until the 7th century CE, some 45km (28 miles) long across the isthmus from the Black Sea to the Sea of Marmara. Also known as the '*Long Walls of Thrace*' or '*Macron Teichos*'.

See 'Antonine wall', 'Çakmak Line', 'Hadrian's wall', 'Theodosian Land walls'.

Anbar. A fortified barracks of the Sasanian Empire (7th -8th century CE, *Iraq*).

Anchoring gad. A rope made of (willow) fibres which was attached to a '*gad*' and an anchoring '*picket*', which was used to secure the '*facines*' to the parapet wall as part of a revetment.

See 'fascine', 'gad', 'revetment'.

Anchoring picket. A wooden stake driven into a parapet, attached to gads to hold '*fascines*', '*gabions*', '*hurdles*' to act as revetments.

See 'fascine', 'gad', 'revetment'.

Anderson hut. Air raid protection for the people of UK (WWII), Named after Sir John Anderson. Consisting of corrugated iron sheets with a semicircular roof, about 2.1 million were distributed starting in 1938 CE, free to families who earned less than £250 but could be purchased for £7. Kit form, half sunk into the ground of your back yard with dirt on the roof (make your own door). Designed to fit four adults and two kids. The shelter was not waterproof and prone to flooding. The alternative shelters were the underground railways and the '*Morrison shelters*'.

See *'flak towers (Germany)', 'Morrison shelter', 'Stanton air raid shelter', 'Winkel shelter'.*

Angle. Angle of the centre of a polygon; the angle of the centre of a polygon obtained by drawing two radii to the ends of the same side.

Angled bastion. An angled bastion provides a minimum target to the front and no dead ground. First true angled bastion in a fortress was built by Giuliano da Sangallo in the early 16th century CE.
Angle of the flank. The angle of the flank is the angle between the curtain and the adjacent flank. Also known as the '*curtain flank*' and the '*flanking angle*'.
Angle of the polygon. The angle of the polygon is formed by the junction of the two adjacent sides of a polygon.
An men. (Ch. *dark door*). A sally port.

See 'postern', 'sally port'.

Annägräben. (G.). Approach trenches perpendicular to the front line.

See '*trench*'.

Annexbatterien. (G. *annex battery*). A covered work provided with artillery that is part of a fort but is built outside the fort itself.

See '*battery*'.

Antardvipa-durga. (Sanskrit *island fort*). An Indian fort surrounded by natural water bodies such as the sea or river.

See 'durga'.

Antemural. (L,) A breastwork or a strong wall provided with towers which project from an outwork.

See 'barbican'.

Antemuralia. (L.) Outworks of a Roman fort like a barbican or breastwork.

See '*barbican*', '*breastwork*', '*outworks*'.

Antesture. A small trench or work formed of palisades.
Anti-aircraft defences.

See 'flak towers', 'barrage balloons'.

Anti-aircraft gun platform. Raised steel platform on roof top in London during WWII, provided with a crewed machine gun and communications for anti-aircraft defence.
Anti-Elephant Spikes. Iron spikes on the front of fort gates to stop the attack of enemy elephants on the gates.

See 'parighas'.

Anti-glider trenches. trenches dug crossing large expanses of open land (England 1940 CE).
Anti-siege tower defences. Buried clay pots or hidden ditches, which could stand the weight of someone walking over them but could fail, upset and topple an advancing siege tower.

See '*siege tower*'.

Anti-personal obstacles. A wide range of obstacles used to impede the infantry.

See 'abatis', 'barbed wire', 'booby trap', 'caltrop', 'cervi', 'chevaux de fise', 'cippe', 'ditch' and 'moat', 'entanglements', 'fakir's bed', 'fraise', 'heishitaji', 'herisson', 'herse', 'hersillon', 'lilia', 'mine', 'mine field', 'monk' and 'wolf's pit', 'punji sticks', 'trap holes', 'trous-de-loup'.

Anti-tank defences / obstacles. Objects and landscape which would obstruct the passage of tanks.

See 'anti-tank obstacles', 'anti-tank mines', 'anti- tank ditches / trenches', 'anti-tank mines', 'dragon's teeth', 'Czech hedgehog', 'tank traps'.

Anti-tank blocks. British WWII concrete blocks were used across the UK as part of the anti-invasion defences of the early 1940s CE and were located above the high-water mark on beaches, on the flanks of approaches to bridges and also flanking the sites of roadblocks. Anti-tank blocks were often employed in front of natural obstacles such as cliffs or rivers) to add to the strength of the feature. The purpose of anti-tank blocks was to seal all beach exits and other vulnerable points in the landscape; halting tanks and other vehicles so they could come under fire from covering positions such as pillboxes or slit trenches. Anti-tank blocks were made in a in a range of shapes and sizes:

Anti-tank buoys were round bottomed cones with a flat truncated top;

Anti-tank cubes were large blocks constructed from reinforced concrete, often constructed on a common foundation, with multiple cubes cast into the same foundation;

Anti-tank cylinders included a countersunk metal eye for the attachment of steel wire rope in the top of the cylinder and a pair of iron cleats cast into the sides of the cylinder;

Anti-tank coffin like the anti-tank pimple, longer along its horizontal axis, with one end lowered along its vertical axis;

Anti-tank pimples are flat-topped pyramids deployed in five rows with the pimples spaced 2.30m (7ft 6in) from their centres;

Anti-tank cone also known as a '*Conical Pimple*';

Anti-tank tetrahedron constructed from concrete tetrahedrons which were more difficult to cast than cylinders and were no more effective. There are some localised variants of the above anti-tank blocks including; a pyramid topped anti-tank cube, a chamfered anti-tank cube and the large cylinder with conical top.

See 'FW3', 'obstacles', 'stop lines'.

Anti-tank ditches, trenches. Excavated ditches to impede the advance of tanks, big enough that a tank would need a bridge, or each side of the ditch demolition by explosives, filled in with earth of fascines. Used by both the Russians and Germans in WWII.

See 'anti-tank obstacles'.

Anti-tank mines. The most common type of anti-tank obstacle which is designed to damage or destroy vehicles like tanks and armoured fighting vehicles. designed to be triggered by a vehicle's weight.

See *'anti- tank obstacles'.*

Anti-tank pimples. Pyramid shaped concrete anti-tank obstacles known as *'dragon's teeth'*. An example is the 570 metres (1870 feet) long stretch on the Isle of Grain, Kent, UK.

See *'dragon's teeth'.*

Anti-tank posts. British WWII reinforced concrete posts set in row about 0.91 m (1 yard) apart used as part of anti-tank defences of an army anti-tank island.

See Army anti-tank island'.

Anti-tank rail. Pointed rails or girders were set in concrete which could damage tank tracks as part of the Maginot Line.

See 'eisenbahnschienen'.

Anti-tank wall. A linear obstacle normally made from reinforced concrete, which was expensive and could be bypassed.

Antonine Wall. A defensive frontier similar to the Limes Germanicus, 63 km (39 miles) long with 19 forts, construction started in 142 CE under the direction of Antonius Pius, which moved the Roman frontier further north of Hadrian's wall, across the narrowest part of Scotland, consisting of turf walls on a stone foundation, with a deep ditch and *'lilia'* on the northern side, from Bridgeness on the Firth of Forth to Old Kilpatrick on the Clyde.

See 'Hadrian's wall', 'limes', 'limes Germanicas', 'limes Raetian'.

Approaches. (Fr.). The trenches which were dug towards a fortress running from the parallel trenches of a besieging force, the trenches were dug in a zig zag pattern so that the occupants were not exposed to enfilade fire from the guns of those besieged in the fortress.

See 'sap', 'trench', 'zig zag'.

Apron.

(1) The portion of the superior slope of a parapet.

(2) The interior slope of a pit designed to protect the rampart against blasts.

(3) The sloping plinth or batter of a fortification's walls and/or towers.

See 'batter', 'plinth', 'talus'.

Apsidal keep. A Welsh keep built on a rectangular plan but with one of its sides rounded. An example is Castle Ewloe, Clwyd, near Chester, built by Llywelyn ap Gruffydd in 1256 CE. Also known as a *'D' shaped keep.*

Arch machicolation. *See 'machicolation', 'pointed arch machicolations'.*

Arc of fire. The scope or number of degrees of the field of fire a projectile weapon has when firing through an aperture in the wall of fortification, such as; an arrow slit, gun port, crenel or an embrasure.

See 'column of fire', 'schussfeld'.

Army anti-tank island (Tank Island). During WWII in 1940 and 1941 CE Britain was expecting a German invasion, *'Tank Islands'* were towns located along the likely route that an invading German army was expected to take, towns like Marlborough were set up with road and railway blocks, defensive pillboxes, strongpoints, with orders to hold the advancing army back for seven to 10 days.

See 'stop lines'.

Army fort (Flak tower). Built by OT (*Germany*) throughout Germany (*WWII*), usually built in pairs, rising up ten to twelve stories, with a battery of guns on the roof, and others for defence, with room enough for crew accommodation, stores, sometimes used as hospitals and air raid shelters.

See 'flak tower'.

Arradah. (Sanskrit). An Indian catapult.

Arresting fort. An isolated and independent fort used to impede an attack.

See 'advanced work', 'fort d'arrét'.

Arris. The sharp edge at the meeting of the salient of a bastion or a spur.

Arrow. An outwork consisting of two parapets fronted by a ditch, situated at the salient angle of the glacis, and was connected to the covert-way by a caponier.

See 'freccia'.

Arrow embrasure (Pointed). Rectangular arrow slit on the outside, rectangular embrasure with pointed top inside.

Arrow headed bastion. A bastion which was used in fortifications to provide flanking as well as covering fire while presenting the enemy with the smallest target possible by the virtue of its shape. Because of its characteristic arrow head shape it could be completely covered by two carefully placed guns in the neighbouring works, in return, a pair of guns in the flank of the bastion could efficiently sweep the straight wall on either side of at least 180 degrees to the front. A fortification employing such bastion in the before mentioned manner is said to have a '*bastioned trace*'.

See 'bastion', 'bastioned trace'.

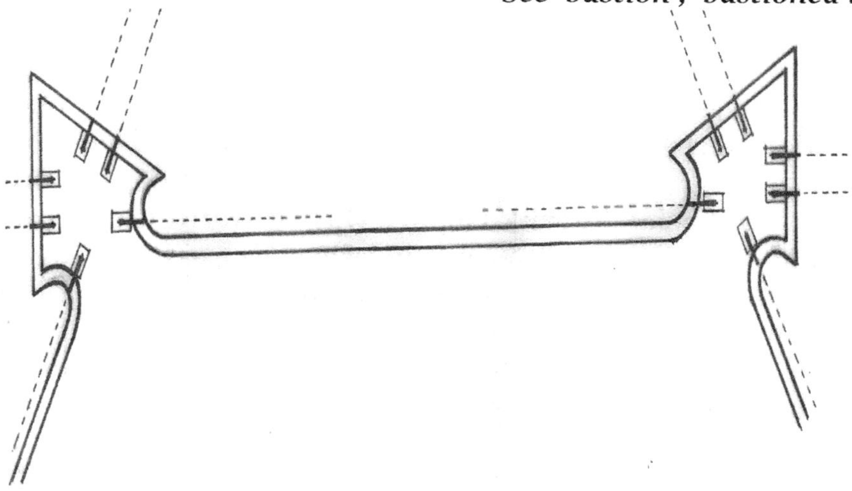

Arrow headed bastion, showing the direction for cannon fire available.

Arrow loop. A narrow opening in the wall of a tower, curtain wall or merlon, which was used by archers to shoot through at the enemy without, with a limited amount of exposure to enemy fire. The arrow loops were carefully spaced in walls to avoid the creation of weak areas which would be subject to damage by siege weapons. The arrow loop was of generally two types. (1) the single vertical slit; and (2) the cross slit, consisting of a vertical slit bisected by a short cross slit.

See 'arrow slit', 'loop', 'oilet'.

Arrow slit. A narrow rectangular opening in the wall of a castle which was used by archers to fire through at the enemy beyond the walls. The earliest form of arrow slit consisted of a plain slit formed by leaving a gap between two adjacent stones of a wall.

See 'arrow loop', 'loop'.

Arrow slit from the inside, the 'embrasure'.

Arsenal. (Ar. *dãr-aş-şinã'a, from dãr 'house' + al- '(of) the' + sinã'a 'art, industry'*). A building for the manufacture and storing of arms, armour and ammunition.

See 'ammunition store', 'enshõgura', 'expense magazine', 'field magazine', 'filling room', 'magasine','polverista', 'principal magazine', 'pulverturm'.

Arthashastra. (Sanskrit). An Indian treatise on a range of topics including military strategy which describes six major types of forts differentiated by their major modes of defences.

See 'regelbau'.

Artillery bastion. A bastion which was defended by cannons.

See 'arrow headed bastion'.

Artillery defences. Fortifications which were defended by artillery.

See 'arrow headed bastion', 'demilune', 'gun platform', 'ravelin'.

Artillery line. The location of the artillery used to bombard the enemy.

Artillery fortress. A fortress designed specially to be defended by artillery, for example the artillery forts of King Henry VII of England, such as Deal Castle (Kent) which was constructed between 1539 and 1540 CE.

Artillery Garden, Artillery Field, Artillery Ground. A large open space either within or just outside a castle or fortified town for defenders to practice archery, musketry or military drill. Early modern period examples include The Artillery Ground in Finsbury, London and Heidelberg Castle (UK), Baden-Württemberg, (Germany).

Artillery outwork.

See 'hornwork', 'ravelin'.

Artillery park. A place for the storage of siege or field artillery, including the cannons, ammunition, carriages, equipment and gun crews.

See 'siege park'.

Artillery platform, Artillery terrace. ***See 'gun platform'.***

Artillery strong point. A work such as a *'battery'*, *'redoubt'* or *'bastion'*, that served to anchor a larger system of fortifications.

See 'bastion', 'battery', 'redoubt'.

Artillery tower. A tower provided with cannon, especially reinforced to take the weight and recoil of the cannon, which were normally situated on the roof, giving them a height advantage over the artillery of the besiegers.

Arx (L.). A castle, citadel, or a keep of a walled enclosure.

Ascensi valli. (L.). The staircases or ramps used to reach the rampart walk of a Roman fort.

Ascent. The stairs or ramp used to access the rampart.

See 'ascensi valli'.

Ashishita sama. (J.). Loopholes at one's feet.

See 'sama'.

Asparagus. ***See 'Rommel's asparagus'.***

Astverhau. (G. *branch entanglement*). A branch entanglement, or tree or branch barrier.

See 'abatic'.

Atlantic pocket. ***See 'festung'.***

Atlantic Wall. The German defences built along the Atlantic ocean coast (from Norway to Spain) designed to stop the allies invading mainland Europe during the Second World War, and was breached by direct assault in 1944 CE at Normandy by the Allied forces on *'D-Day'*. Consisting of beach obstacles, bunkers, casemates, gun emplacements and other fortifications, designed and built under the direction of the Organisation Todt (OT).

See *'Festung Norwegen', 'Südwall'*.

Atalaya. A Spanish watch tower, common in northern Spain. Examples are in Torrelodones and El Berrueco.

***See 'almenara'*.**

Attataka. (Sanskrit). A tower on a parapet, built in a round, square or rectangular shape.

Aurelian Wall. A defensive wall that encircled Rome, Instigated by the Roman Emperor Lucias Domitius Aurelain, 271–275 CE. The wall encircling was 18 km long, took 10 years to build, up to 4 metres thick, with 381 towers (about a tower every 30 metres (98 feet)). Wallwalk at 6 metres (20 foot) high, towers at 4.5 metres (15 feet) higher than adjacent walls. Refurbishments were carried out over time until the Sack of Rome by Alaric and the Goths (410 CE).

See '*Servian wall*'.

Auxiliary fort. See 'castra'.

Avanceé. A '*place of arms*' located where the road leaves the '*covertway*' after passing through a gate.

Avant-cuirasse. (Fr. *forward armour*). Glacis armour, armour that protects and armoured turret.

Avant poste. (Fr. *outpost*). A small temporary or permanent place housing soldiers, forwards of a larger fortification, border or disputed region between states, used to detect incursions or attacks.

Avans mur. (Fr. *front wall*). **See '*antemural*'.**

Azuchi umadashi. (J.). An earthen embankment ('*doi*') or wall built in front of a castle's gate ('*koguchi*') to act as defensive barrier ('umadashi') to defend an attack on the gate, acting like a '*barbican*'. The barrier could be surrounded by a moat ('*hori*') and could be rounded ('*maru umadashi*') or square ('*kaku umadashi*') in shape.

See '*barbican*', '*doi*', '*hori*', 'kaku umadashi', '*koguchi*', '*maru umadashi*', '*umadashi*'.

B.

Back ditch. A ditch dug behind a parapet. Also known as an *'interior ditch'*.

See *'ditch'*.

Bail.

(1) The outer line of a fortification.

(2) The wall surrounding the outer courtyard of a castle, or the wall separating the courtyards of a castle.

(3) The courtyard or bailey of a castle.

See 'bailey'.

Bailey. (Fr. *baille*, palisade, enclosure; from L. *ballium*). The enclosed space or court between the keep and the curtain wall or palisade surrounding it. In *'motte and bailey'* castles, the bailey was used as a protective enclosure for the domestic and working buildings associated with the donjon. In later fortifications the bailey was kept clear so as to remove any cover for attackers crossing the bailey to the next line of defence.

See *'motte and bailey castle'*.

Bailey walk. *See 'allure', 'wall walk'*.

Bailey wall. *See 'curtain wall'*.

Bai lu wu. (Ch. *white dew hut*). A small domed shaped hut made from bamboo or willow, only large enough for a single person built on top of a *'zhan peng'*. Used as a temporary shelter for a watchman and storage for firefighting equipment.

Baliq. Turkish small fortified town.

Ballistra. A siege engine with two arms held under tension by scions of ropes, sat on a stand, later a wheeled frame, the string connecting the ends of both arms was drawn back by a windlass with wooden levers, a large arrow or other missile was loaded and the string released propelling the missile towards the enemy. An example is the story of the three or more Vikings struck by the same missile during the siege of Paris in 885 CE.

Balistraria. A crossbow loophole consisting of a small cross in the wall of a fortification, used by crossbowmen to defend their position. Also known as an *'arbalestina'*.

See 'crosslet slit', 'loop'.

Ballistarium. A platform used to support an onager or ballista which was used to defend a Roman fort.
Baltic defence line. Linear defences built to defend the countries of Estonia, Latvia and Lithuania, about 1,360 km (845 miles) long, to defend against Russian aggression in 2025 CE after the invasion of Ukraine, consisting of ditches and ramparts, dragon's teeth, hedgehogs and razor wire, and defended river crossings.

See 'linear fortifications'.

Banquette. (Teut. *bench*). The platform of earth within the parapet, high enough to enable defenders to fire over the crest of the parapet while standing.
Banquette slope. The earthen slope behind the banquette leading to the '*covered way*' or '*terraplein*', where riflemen could find cover behind the 'parapet' to reload.

See '*banquette*'.

Bansho. (J.). Castle guardhouse.
Barbacan, Barbican. (O.Fr. *barbacane*). An outwork consisting of an outer bailey provided with flanking towers, which was used to defend the main gate of a fortified town, castle or the approaches of a bridge situated before a fortification. Normally projecting from the main entrance, the shape of which was dictated by the site and the design used. One type consisted of two walls parallel to each other projecting

from the main gateway with another gateway situated at the other end; the first gate was flanked by towers which were provided with battlements, arrow slits and machicolations. Also known as a '*forecourt*'.

See 'forework', 'umadashi'.

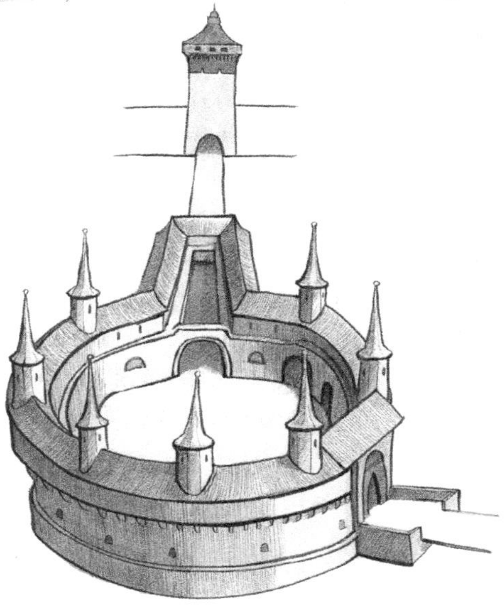

Barbican, Kraków, Poland.

See '*dwingel*'.

Barbed wire. Fencing wire with sharp protrusions or barbs, developed in the USA, used to fence in livestock. Used to control access to large areas (i.e. trench warfare). Entanglements of barbed wire could be multiple rows deep. Breaks in the barbed wire were placed at certain intervals which allowed troops to access '*no man's land*'.

See 'entanglements'.

Barbed wire entanglements. Used to control access at frontiers and defend ditches, trenches and other fortifications.

See 'barbed wire', 'entanglement', 'high and low wire entanglements', 'spider wire entanglements'.

Barbette, Barbe, Barbet, Barquette. (L. *barba*, beard). An earthen terrace or platform situated inside the parapet or a rampart, upon which cannons were mounted so that they could be fired over a wall rather than through a gun port. A battery in this situation is called a '*battery en barbe*' (or '*barbet*').

Barbkin, Barmkin (Scots).

(1) A walled enclosure surrounding a pele tower.

(2) The courtyard attached to a tower house.

See 'peel tower', 'pele tower', 'tower house'.

Bar-lev-line. A defensive line constructed by the Israelis along the eastern bank of the Suez Canal in 1967 CE, 150 km (93 miles) long, with forts and strongpoints. built along the line, but was not effective in stopping the Egyptians in 1973 CE. The strong points included; bunkers, troop shelters, positions, firing positions for tanks, protected by barbed wire, minefields and pipes to spread flammable liquid onto the surface of the Suez Canal.

Barquette. ***See 'barbette'.***

Barrage balloon. A large tethered un-crewed balloon used in WWI and WWII (UK) to defend ground targets against aircraft attack by raising aloft steel cables, in some cases steel netting, which would be hazardous to aircraft. The barrage balloons were tethered to concrete anchors or launched using specialist trucks.

Barrage balloons over London, UK.

Barricades. Temporary barriers that are only useful against small arms fire, closing off an area, using whatever comes to hand, such as vehicles, boxes and barrels.

Barricata. (It. *barricade*). A bunker with a fixed 50 calibre machine gun.

Barrier. A timber temporary obstruction used for protection against small arms fire.

Barrière. The network of twenty five fortified cities in Belgium used to create a buffer zone to prevent future French aggression after the Napoleonic Wars (1803 – 1815 CE), associated with the Duke of Wellington, thus it was also known as a '*Wellington Barrier*'.

Barrière de fer. (Fr. Iron Barrier). Consisting of clusters and lines of detached polygonal forts to defend Paris, Lyon and the eastern border of France (about 2044 km (1270 miles)) of border from Dunkirchen to

Toulon, from attacks by Prussia, constructed between 1874 and 1914 CE, designed by General Séré de Rivières.

See 'feste', 'polygonal fort'.

Barrier forts, Barrier works.

(1) Detached works designed to close a defile, mountain passes and waterways, usually sited on the flanks of a field army, consisting of a line of '*supporting points*', used to hold up an enemy's advance.

(2) A mutually supportive work which forms a line of defence.

See 'barrier'.

Bartisan, Bartizan. A turret which projects at an angle from a tower, a parapet or near a gateway. Used as a watch tower or a defensive position by utilising flanking fire. Bartizans were also used as strengthening buttresses or simply as decoration on later castles or castellated residences.

See 'échauguette', 'flanking turret', 'machicolated turret', 'machicoulis', 'tourelle'.

Bartizan, Fort George, Inverness, Scotland.

Bartizaned. Provided with bartizans.

Bascule bridge. (Fr. *bascule*, see-saw). A kind of drawbridge used to defend a gateway. The bridge was moved using a counterpoise system, where a counter weight was moved into a pit, thus bringing the bridge up to fill the gateway.

See '*bridge pit*', '*drawbridge*', 'gaffs', 'gatehouse'.

Base. (L. *bassus*, short). The line connecting the salient angles of a bastioned front.

Base aèroterrestre. (Fr. *air-ground base*). Using air power to set up and supply a base in enemy territory. Initially successfully used by the French during the Vietnam War (1946 – 1954 CE) but ultimately failed.

See 'fire support base'.

Bashura. (Ar.). The bent gates used in Baghdad in the 8th century CE.

See 'bent entrance'.

Basson. (Langue d'oc). A battering ram.

Bastajas. The Polish fortification consisted of a low, wide, round tower set out from a wall to give an ample platform for artillery, which later evolved with gun ports, artillery embrasures. Some Bastajas reached five stories high. Examples can be found in Krakow, Wroclaw, and Warsaw (Poland), Wilno (Russia).

See '*barbican*'.

Bastei. (It. *bastia*, bastion) Dürer's '*bastei*' or tower was armed with long range cannon, caponiers and casements with gun ports for close defence, the casemates had flues to take away the smoke to aid with ventilation and observation.

Bastide. (Old Fr. *fortification, fortified place*).

(1) A type of small fort.

(2) A fortified town which was common to the south-west of France, dating from the 12th and 13th centuries CE.

Bastille. (Old Fr. *baastel, basstel,* fortification; fortified tower). Originally, a kind of siege tower, later the term was used to refer to a tower which had the curtain walls and the flanking towers at the same height, permitting the rapid deployment of troops and cannon for the defence of any threatened sector. The lowering of the towers to the same height as the curtain wall was a development brought about by the effect of cannon fire on more lofty towers.

Bastion. (L. *bastia*, build). A work projecting from the curtain wall of a fortification which commanded the foreground and the outworks. Designed to provide flanking fire to the adjacent curtains and bastion.

Bastion has been used to refer to the flanking towers of a castle as well as the arrow headed bastions of the Italian bastion trace.

See 'arrow headed bastion', 'curtain tower', 'mural tower'.

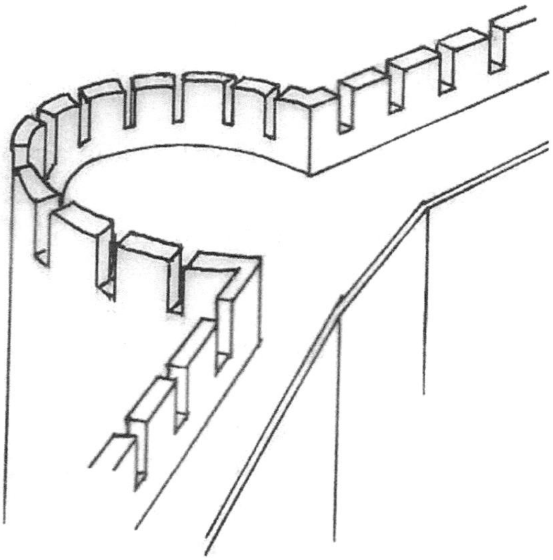

Bastioned curtain. ***See 'reinforced order'.***

Bastion spur. A sharp piece of masonry constructed at the point of a bastion.

See 'spur'.

Bastion systems. Various systems of defensive works incorporating bastions, from the Italian (mid-15th century CE) to the polygonal of the 18th century CE, including:

the old Italian system;
the new Italian system;
the French system, 1st Vauban;
the French, system, 2nd Vauban;
the French system, 3rd Vauban;
the old Dutch system;
the new Dutch system (Coehoorn);
the tenaille-trace, tenaille-system;
the polygonal system (-trace) or polygonal fortification.

See *'bastion', 'bastion fort', 'bastion trace', 'polygonal trace', 'tenaille trace', 'trace', 'trace italienne'.*

Bastioned flank. The salient angle of a bastion and the opposite flank.

Bastioned trace. The geometrical system of arrow headed bastions and ramparts, which were used to defend fortifications dating from the 15th century CE.

See 'arrow headed bastion', 'bastion'.

Bastioned tower. A pentagonal shaped tower two levelled casemated tower, providing protection to flanks and the main wall, with a top floor built as an artillery platform, which could be provided with a roof, which functioned as a bastion in Vauban's second system, often provided with a counterguard in fronts, useful for difficult terrain, like at Besançon (France).

See '*Vauban*'.

Bastioned tower, Besançon, France.

Bastionet, Bastionettes. A small type of bastion attached to the salient angle of a bastion, so as to increase the amount of flanking fire which could be used to flank the ditch but is not a principal defensive element.

See 'bastion'.

Bastion fort. A style of early '*gunpowder*' period fortification, mid-15th century CE in Italy, combining ravelins and other outworks, looking like a '*star fort*'. Also known as '*trace italienne*' (Italian outline).

Bastle. (Fr. *bastille*). A defensible stone farmstead unique to Britain, dating from the 16th to the early 17th century CE. Designed to defend the occupants and their livestock from raiders. A bastle consisted of two storeys with a windowless ground floor which was used to house the livestock, its entrance was secured from within by a heavily barred door and a trap door to the upper storey; the upper storey itself had its few small windows secured by iron bars, and its narrow door was situated high above ground level and was gained by a removable ladder. Bastles tended to be built in clusters for mutual support.

Bastle house. A larger version of the bastle.

See 'bastle'.

Batardeau. (Fr. *cofferdam*). A wall or coffer dam which traversed a ditch of a fortification, equipped with a sluice gate which was used to regulate the height of the water in the ditch, but was rendered impassable because of its knife edge apex.

See 'coffer', *'dame', 'monk'.*

Batter, battering. (L. *batu*, beat).

(1) A wall with a receding slope from the ground upwards, narrowing at the top, is said to be battered.

See 'plinth', 'spur', 'talus'.

(2) To use a siege engine or artillery to batter or strike repeatedly against a fortification's wall or gate to make a breach.

Battered base, Battered plinth. The projection at the base of a wall which sloped outwards. Also known as *'battering'*.

See 'plinth', 'spur', 'talus'.

Battered wall. A wall of a fortification provided with battering.

See 'plinth', 'spur', 'talus'.

Battering. *See 'batter'.*

Battering ram. A tool consisting of a large piece of timber, sometimes with a metal head, used to breach a part of a fortification, be it wall, tower or gate. Could be suspended below a siege tower or a penthouse to protect the soldiers swinging the ram while the weight is taken by the structure above.

See 'penthouse', 'siege tower'.

Batteries. Any place where guns or artillery such as mortars are set up.

Battery. A work consisting of an epaulement or breastwork which was used to protect a gun or mortar emplacement, when used for guns, embrasures were made in the parapet so that the guns could be fired through them.

Battery fascines. Fascines used for revetments, made in bundles 5.5m long and 25cm (18 feet long by 10 inches) in diameter. Also known as '*long fascines*' or '*saucissons*'.

See '*fascines*', '*revetment*', '*saucissons*'.

Battery tower. A defensive tower added to the outer defences of a castle, dating from the 16th century CE onwards, corresponding with the advent of cannons. Usually a round tower with multiple embrasures on multiple levels facing all the approaches.

See '*battery*'.

Battlement. (O.Fr. *batailler*, movable defences). The upper part of a fortification's wall from which defenders protected the wall top from attackers and found protection from incoming missiles. The battlement or parapet was usually provided with crenels and merlons, the crenels were the openings and the merlons were the solid uprights. This arrangement allowed the defenders to fire upon attackers through the crenels while obtaining some protection from the returned enemy fire behind the merlons.

See 'merlon', 'parapet'.

Battlemented. Provided with a battlement or parapet, which usually were divided at regular intervals by crenels and merlons. Also known as crenelated or embattled.

See 'battlement'.

Baulks. A squared off timber beams used to support the planks of a bridge crossing a ditch.

Baumbeobachter (G. *tree watcher*).

See *'camouflaged tree'.*

Baumstämmen. (G. *logs*). A defence of felled interlocked trees.

See 'abatic', 'astverhau'.

Bawn. (Ir. *báhun*, enclosure).

(1) A fortification which was used in Ireland to defend a house, or a cattle enclosure, which consisted of a surrounding stone wall.

(2) An Irish word used to refer to a curtain wall.

Bay. A section of trench that lies between two adjacent traverses and was used as a passing place.

See *'trench'.*

Beach head. A temporary line created when military landing craft reach a beach, creates space and begins to defend the area as other reinforcements arrive (i.e. the beaches of Normandy, France, WWII).

See *'lodgement'.*

Beach obstacles. Various devices used to hamper beach landings.

See 'admiralty scaffolding', 'Atlantic Wall', 'belgian gate', 'hemmbalk', 'hemmkurve', 'küstenvorfeldsperren', 'Rommel's asparagus', 'tetrahedron'.

Beacon. (Old English, bêacn, sign, portent). A fire set a light on top of a hill, pole or building, to warn of impending attack by an enemy. Examples are, England 1330's CE, beacons were set up along the coast and inland against attack by France during the *'Hundred years' war'*. In the 16th century CE, England set up a new system of coast beacons to warn of the attack of the Spanish Armada. In the 1600's CE rivalry between the Scottish Armstrongs and the English Grahams saw towers provided with beacons for warning of attacks.

Beak, Beaked projection. See *'en bec'.*

Bee hives. Used by defenders to deter attackers. Examples are: the defenders of Chester threw beehives at the Vikings in 918 CE; during the First Crusade the Moslems of Ma'arrat used a ballista to hurl beehives at the Crusaders.

Befestigungsanlage. (G.). Fortifications.

Belagerungsburg. (G. *siege castle*).

See *'gegenburg', 'counter castle', 'malvoisin'.*

Belfry, Beffroy. (Middle English, *berfrey*). A wheeled siege tower which could be provided with a ram and defended by archers, used to attach a fortification's walls and provide access to the wall top via a drawbridge. The earliest depiction of a siege tower is from a 700 BCE stone slab from the palace of Assur-Nasir-pal at Nimrod in Assyria. Alexander the Great used prefab wooden towers during his conquests. The Avers used wooden towers when they attacked Constantinople in 626 CE. The concept was last used in 1945 CE to take Fort Drum in Manila bay during WWII, in the form of modified landing craft with a bridge to allow troops to access the top of the fort.

See 'siege tower'.

Belgian Gate. A mobile anti-tank obstacle consisting of a heavy steel fence on rollers, designed by the French, also used as barricade for bridges, roads and fort entries, repurposed by Germans as a beach obstacle. Also known as '*cointet element*', or more simply '*c-element*'.

See 'obstacle'.

Belmont anti-tank ditch. An Australian anti-tank barrier built in 1942 CE in Belmont (New South Wales) to counter a possible Japanese landing, consisting of an anti-tank ditch with barbed wire, '*dragon's teeth*' and '*dumble tank stops*'.

See 'dumber tank stops'.

Belt fortress. See '*gürtelfestung*'.

Beluardetto. A small bastion used on a fausse-braye.

See '*fausse-braye*'.

Benes Lines. A line of defences consisting of forts, bunkers, artillery emplacements and obstacles built by Czechoslovakia along their border with Germany (815 kilometres (506 miles)), started in 1935 CE, based on the French Maginot Line, and named after the Czechoslovakian President Benes.

Benin Moat (Edo. *Iyanuwo*), **Benin Iya** ('*Walls of Benin*'). The moat and earthen rampart system that protected the city of Benin (the principal city of the Edo (Bini) kingdom of Benin, which controlled a large part of west Africa (Nigeria), dating from the 8th to 19th Century CE. The '*Walls of Benin*' consisted of outer earthworks for the city and inner earthworks for the palace, with towers and gates, a total of about 1,200 kilometers (746 miles) in length, the inner earthworks were about 10 kilometres (6.2 miles) long. The City of Benin was largely destroyed by the British invasion of 1897 CE. Also known as '*Iya*' or '*Iyanuwo*'.

Bent entrance. An entrance of a fortification involving one or more sharp changes in direction. If an enemy force gained entry they had to then turn because of the shape of the entry and this exposed their unshielded side to the fire of the defenders.

See 'clavicula', 'masugata mon'.

Beobachtungsglocken. (G. *observation bell*). Observation cloche with small openings.

Beobachtungsschnecke. (G. *observation snail*). A German WWI observation post of a *'feste'*, called a snail because of the shape of its plan, armoured with two layers of steel with a layer of sand in between, or some were made in concrete.

See 'cloche', 'feste', 'tobruk'.

Beobachtrungsposten, Beobachtungsstände. (G. *observation stand*). A German observation and fire control cupola or cloche with small openings, usually fixed and accessible from the bunker. Also known as a *'kleinglocke'*.

Bereitschaftsraum. (G. *ready room*). A room providing housing for troops in German bunkers of the 1st and 2nd WW, provided with standardised furniture (tables, chairs, cupboards and bunk beds).

Bergfried.

(1) A single defensive tower characteristic of German speaking lands, the chief function of which was as a watch tower and as a final refuge, rarely were they used as permanent living quarters. The entrance was situated on the first floor.

(2) A watch tower which was used to cover the main lines of approach to a castle, normally associated with a *'hohenburg'*.

See *'hohenburg'*.

Bering vorburg. (G.). Outer bailey.

Bernhardt Line. ***See 'Reinhard line'.***

Berquil. (Ar.). Large outdoor water reservoir found in Crusader and Arab castles.

See 'cistern'.

Berlin wall. A border wall surrounding West Berlin from the German Democratic Republic (GDR), built in 1961 CE, removed in 1994 CE, consisting of guard towers, bunkers, concrete walls, anti-vehicle trenches and barbed wire entanglements, 155 km (96 miles) long.

Berm. (Fr. *berme).* A narrow path that runs between the ditch and the parapet, and when it is only made of turf its purpose is to prevent soil falling into the ditch.

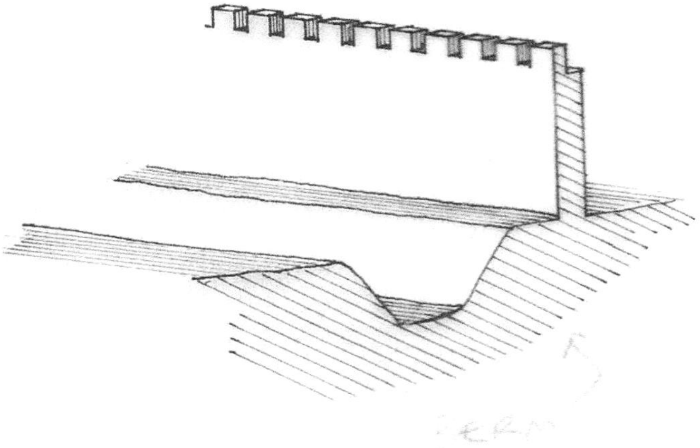

Besiege. To attack, invest, or to lay siege to a place or fortification.

See 'mine', 'sap'.

Bettung. (G.). Firing position, usually concrete.

Biehler fort. A German polygonal fort, which included dry ditches with a caponier, barracks in the gorge, numerous artillery emplacements defended by traverses. About seventy forts were built to protect various German cities like Cologne, Spandau, Ulm, during 1870 – 1890 CE. Also known as '*Biehlersches Einheitsfort*' or '*Schemafort*'.

B-Kazemat. A type of Dutch reinforced concrete casemate, a 1930's CE part of the pre WWII defensive program, defended by a machine gun or anti-tank weapons and a small crew, with an armoured door on the flank to the embrasure. Left orientation, used to defend river crossings and key strong points. Right orientation used along defensive barriers like the '*Grebbe line*'.

See 'casemate'.

Blast pen. A place surrounded by two or three embankments, used for the storage of aircraft, vehicles and other equipment, to provide protection from enemy attack.

See 'hardened aircraft shelter'.

Blast wall. A wall used to protect structures or vulnerable buildings and the people inside from the effects of explosions (i.e. '*Hesco bastions*' used in Afghanistan, or entrances to bunkers and pillboxes).

See 'bunker', 'Hesco bastion', 'pillbox'.

Blind. A piece of wood placed over a trench to support a covering of hurdles.

See *'gabion', 'hurdle'.*

Blindage. (Fr. *shielding*).
(1) A structure over a trench made of logs (and other materials) protecting the soldiers in the trench.
(2) A bunker made for earth and timber (and other materials) used to provide structure to protect headquarters, field hospitals and rest areas.

Blind arcade / arches. A row of arches or an arcade built on the inside of a wall to add strength or to increase the width of the wall walk only open on the back.

Blindes. Timber planks placed over a trench to support sandbags and other material, to provide protection to the occupants of the trench.

Blind gun port. A gun port consisting of a single round hole without a sighting slit, behind this was a much larger rectangular space widening towards the inside wall, big enough to operate the gun. The gun ports were designed to allow defenders to operate a small gun mounted on a stand, or clamped to a wooden bed, but the space soon filled up with smoke after a few firings because of the lack of ventilation.

See 'gun loop', 'gun port', 'loop'.

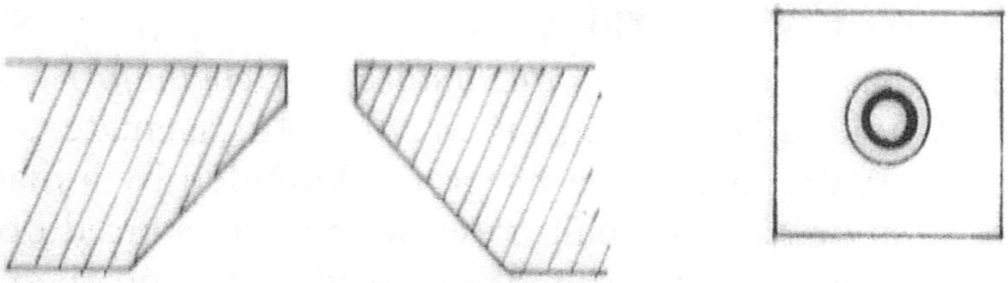

Blind gun port, plan view and front view.

Blind machicolations. Machicolations which were built without openings which were used as decoration, machicolations became defunct due to the introduction and perfection of artillery in siege craft, dating from the late 14th century CE.

See ***'bracket', 'bracketing', 'false machicolation', 'mock machicolation'.***
Blind spot. ***See 'dead angle', 'dead ground', 'flanking'.***
Blockade. A siege tactic to block access to a place which restricts access to food, water and relief forces.

See Appendix 3. Siege Warfare - The art of Offence and Defence.

Bloc. A reinforced bunker of the Maginot Line which mainly consisted of; armed combat blocs, defended entrance blocs, and observation blocs.

See 'bunker', 'Maginot Line', 'ouvrage'.

Block-castle. A castle with walls and towers built to the same height, with continuous wall walks, and a small bailey in the centre, dating to the 14th century CE (i.e. The Bastille in Paris, France).
Block house.
(1) A fortification used for seaward defence provided with; shot deflecting battlements, hand gun ports and a single embrasure for a long-range cannon, used during the 16th century CE.
(2) Small detached fort at a strategic point, later just referring to a wooden structure. An example of a block house can be found in the centre of the New York Central Park, dating from 1812 CE.

See *'Boer War blockhouses'.*

Blockhaus. A stand alone infantry combat position armed with machine guns.

See 'blockhouse', 'bunker', 'ouvrage', 'pillbox'.

Blunt. To remove a salient angle by the construction of a parapet between two salient works.

See 'pan coupé'.

Body of the place. Either the buildings of a fortification or the works surrounding them, but generally when the body of the place is constructed it refers to fortifying the place by enclosing the fortification with curtain walls and bastions.
Boiling oil holes. ***See 'murder holes', 'meurtriéres'.***
Bombproof. A structure specifically designed to provide protection against heavy artillery fire.
Bonnet. A small triangular work consisting of two faces which was situated in front and parallel to the salient angle of a bastion or a ravelin, which provided additional defence against enfilade fire.

See *'counterscarp bonnet', 'rear bonnet'.*

Bonnet de Pretre. (Fr. *priest cap*). A defensive work that looks like a *'tenaille'.*

See *'priest's cap', 'tenaille'.*

Booby trap. [A trap to catch a bird called a '*booby*']. Traps and devices used for anti-personnel warfare, such as the ones used by the Viet Cong during the Vietnam War. Punji sticks of bamboo or nails set up in holes in the ground which would pierce boots and legs; land mines placed about obstacles that block paths; grenades with their pins pulled by a trip wire across a path.

See '*trip wire*'.

Boer War blockhouses. During the Boer War [1899-1902 CE] in South Africa there were a range of styles of blockhouses used to control territory and protect the railways in South Africa against attack by the Boers, in conjunction with barbed wire entanglements and connected by telephone. There were over 8,000 blockhouses built along 6,000 km (3728 miles) of defensive lines, dividing South Africa up into controllable areas. About 50,000 soldiers were deployed and the whole system of defences cost, at the time, £1,000,000. Initially the blockhouses were constructed of masonry and took about three months to build, costing up to £1,000 each, thus the design was changed to a prefabricated structure.

The barbed wire entanglements between blockhouses were provided with bells and various noisy items like tin cans placed on the wire to sound an alarm when being tampered with.

Masonry blockhouses of the Boer war were mostly built to the standard three-storied blockhouse design by Major-General E. Wood (Chief Engineer) using mortared stonework or reinforced concrete. The masonry blockhouses were built at bridges, stations and towns, provided with: the ground floor used for storage, living quarters on the 1st floor, the 2nd storey for observation. Two steel galleries '*machicoulis*' were sited at diagonally opposite corners to allow flanking fire along the walls in case of an attack. A ladder (or ladders) was used for access to the block house and between floors. The 441 masonry blockhouses built for the Boer War came in a number of shapes including hexagonal and octagonal. A cheaper prefabricated corrugated iron blockhouse was also developed.

See '*rice blockhouse*'.

Boom defences. A device to stop access to a waterway or harbour consisting of a chain and or floating elements. Examples are: Constantinople 10th to the 12th century CE, Portsmouth (UK, 1420-1422 CE, Southampton Water (UK, 1890 CE, consisting of Hulks attached stem to stern), Scapaflow's Great Harbour (UK, 1915 CE).

Bore. (L. *terebra*, drill). A strong pole with a sharp iron head used by sappers to attack the base of a fortification's wall.

See 'mouse'.

Bori. (J.) Trenches used to defend Japanese castles.

See 'hakobori', *'horikiri', 'kenukibori', 'shojibori', 'tatebori', 'unebori', 'unejotatebori', 'yagenbori'.*

Borg. (N.). A tower, fortress or castle.

Borj, Burj. (Ar.) A fort or tower, found in northern Africa of various designs (i.e. Borj Nord in Morocco, or Borj El Kebir in Tunisia).

Borogata. (J. *watch tower*). An early type of '*tenshu*' which was built on an irregular shaped stone foundation.

See *'sotogata', 'tenshu'.*

Boroshiki tenshu. (J.). The earliest form of Japanese keep or '*tenshu*', consisting of a multi-storied timber framed tower with a belvedere or observation tower set on top of the roof ridge.

See *'borogata'*, 'dokuritzushiki', *'fukugôshiki tenshu', 'renketsushiki tenshu', 'reritsushiki tenshu', 'teiritsushiki tenshu'.*

Borough. *See 'burh'.*

Boteras. (O. Fr.). A type of buttress.

Bottom. A round disk in which holes had been drilled or punched, rods were then inserted to form a basket called a gabion, these gabions when filled with earth were used in the construction of earthworks.

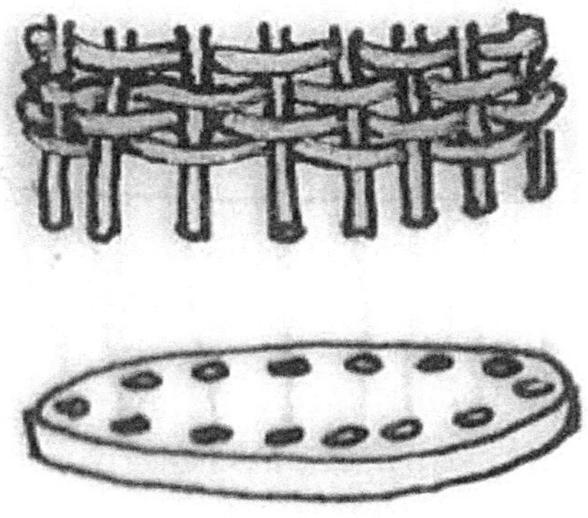

See *'gabion'.*

Bottom of the ditch. The flat space at the base of the ditch that separated the '*scarp*' and the '*counterscarp*'.

See *'counterscarp', 'ditch', 'scarp'.*

Box machicolation. A machicolation developed in the Middle East around the 12th century CE. Each machicolation was built out from the curtain wall separately from its neighbour, and was enclosed at the top, the top was below the level of the wall walk above. This type of machicolation was also at times provided with an arrow slit.

See 'machicolation'.

Katiebell © 2024

Boyau, Boyeaux. (Fr.). A trench used by a besieging force, also known as a zig zag, used for the purpose of communication in such situations as; a trench from the rear of a battery to the magazine, or a trench from the approaches towards a besieged place.

See 'approaches', 'sap', 'trench', 'zig zag'.

Box. A British field fortification constructed and used in North Africa during WWII, defended by mines, trenches and saps, foxholes, artillery emplacements, dugouts, storages, a kitchen, and a medical facility. The '*Baggush Box*' was built in the Western Desert near Maaten Baggush, Egypt.

Bracket, Bracketing.

(1) A wooden or masonry projection near the top of a wall which was used to support a '*hoarding*' or a '*bretache*'.

See 'console', 'corbel'.

(2) Machicolations became redundant due to the effect of artillery on castles, so the openings were closed but the brackets remained, later castellated residences used them as decoration.

See 'blind, false, mock machicolations'.

Braie. (L. *brãca*, breeches). A low defensive platform which was used to impede an enemy's assault on the lower walls of a fortification.

See 'bray', 'fausse braie'.

Branch gallery. A small tunnel dug from a *'listening gallery'* to intercept enemy mines.

See *'mine'*.

Brandkugel. (G. *fire ball*). An incendiary projectile consisting of an iron sphere filled with incendiary material thrown at enemies buildings.

Brattice, Brattish, Brettice. (O.Fr. *breteshe*, L. *bretachia*).

(1) A small stone gallery built out from a castle's parapet or wall on corbels but lacking foot boards. Used to reduce the dead ground at the base of the wall by allowing defenders to drop missiles through the hole onto the enemy below. A very similar construction used as a latrine and usually suspended over the moat was known as a *'garderobe'*.

(2) A wooden gallery or balcony built out from a parapet of a castle, used to reduce the dead ground at the base of the walls by allowing the defenders to drop missiles through holes in the floor and fire arrows at the enemy attacking the walls. Later, such wooden galleries were replaced by stone machicolations because of their susceptibility to fire.

See *'hoard'*.

Brattice work. The brattice or hoarding which was used to defend the area at the base of a castle's walls.

See *'brattice'*.

Bray, Braye. Braga. A low wall constructed in front of the ramparts, inside the ditch and separate from both the *'scarp'* and *'counterscarp walls'*, which was designed to keep the enemy at a distance from the ramparts.

See *'braie'*.

Break joints. The overlapped joints of a *'sod revetment'*, which increases the strength of the sod wall.

See *'revetment'*.

Breach. (It. *breccia*). A break made in a fortification's defences by an enemy's artillery, mine or other siege technique or equipment.

See Appendix 3. Siege Warfare - The art of Offence and Defence.

Break ground. To start a siege by digging trenches and other earthworks.

Break joints. The overlapped joints of a '*sod revetment*', which increases the strength of the sod wall.

See '*revetment*'.

Breast height. The inner slope of a rampart, parapet or wall, at chest height, so rifles or muskets could be fired at the enemy with some protection.

Breastwork. An earthwork thrown up to breast height which provided protection to defenders firing over the crest of the work while in a standing position.

Bremer wall. A portable wall made of steel reinforced concrete, used as a '*blast wall*' in Iraq and Afghanistan. Similar to a '*Jersey*' traffic barrier but taller. Originally designed by the Israelis. It was called a '*T-wall*' because if inverted it formed the letter '*T*'. Also known as an '*Alaska barrier*', '*T-wall*', '*Texas barrier*'.

See '*blast wall*'.

Breteche, Brettice. See '*brattice*'.

British hardened field defences. British WWII small emplacements built under the '*FW3*' to defend Britain from invasion by Germany consisted of mainly '*pillboxes*'.

See 'allan williams turret', 'cantilevered pillbox', 'dover quad', 'eared pillbox, 'eastern command variants', 'essex lozenge pillbox', 'field gun emplacements', 'FW3', 'Lincolnshire three-bay', 'lozenge pillbox' 'northumberland 'D' Type', 'pentagonal pillbox', 'pillbox', 'pill box 22, 23, 24, 25, 26, 27, 28', 'ROF Type', 'section post/seagull trench', suffolk square', 'vickers machine gun emplacement'.

Briton machicolation. A machicolation used during the 14th century CE which was developed in Brittany, built out on corbels which were sharpened to a point to give defenders a wider field of fire than the machicolations built on the straight corbels.

See '*machicolation*'.

Bridge castle. See '*Brückenburg*'.

Bridge (flying). See '*flying bridge*', '*motte*'.

Bridge (fortified). See '*bridge castle*', '*brückenburg*', '*fortified bridge*'.

Bridge head. A work made of one or more '*bastions*' or '*redans*', used to protect a bridge from enemy fire. Also known as '*tete-du-pont*'.

See '*tete-du-pont*'.

Bridge pit. A stone lined pit located before a gatehouse which was covered by the drawbridge when in the open position, and when the drawbridge was closed the pit was exposed and formed an impediment to attackers.

See '*bascule bridge*', '*drawbridge*'.

Brigola, Brigole. (Sp.). Likely different names for a '*trebuchet*'.

Brisure. (Fr. *briser*, to break).

(1) A rampart or parapet or part of either that does not follow the general direction of a work, of which it is a part.

(2) Where the line of the curtain wall is broken to allow an expansion of the space for the guns in the flank of a bastion.

Broch, Brogh, Brough. (N. *borg*. fortress). A dry stone circular tower of the late Iron Age, common in the north of Scotland, very rare in the south. The surrounding was built solid at the ground level, while higher up the wall it divided into a number of chambers and galleries. The small doorway through the base of the wall was easily defensible. The design of the broch showed a need to build upwards rather than outwards, so as to reduce the size of the perimeter so that fewer people were needed to defend it. Note: It appears in some text that '*dun*' and '*broch*' are substituted.

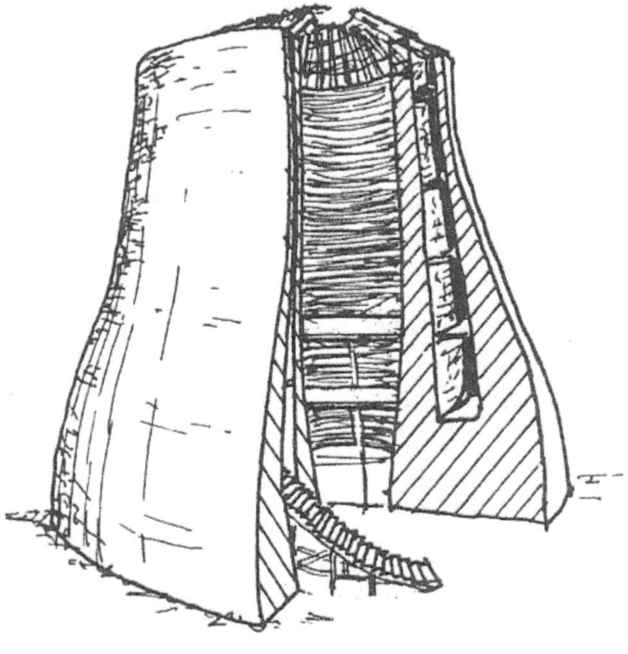

Cut through the Broch of Mousa, showing the various galleries, stairway inside walls, and internal floors and roof.

See '*dun*', '*pictish tower*'.

Brückenburg. (G. *bridge castle*). A type of German castle built to provide control of a bridge over a river, could be used to collect tolls, not necessarily built as part of the bridge as in a '*fortified bridge*'.

See *'fortified bridge'*.

Bruno's spiral. A spiral type of defensive barbed wire where the barbed is wound around a cylinder. Used as part of fences.

See *'barbed wire'*, *'entanglement'*.

Brustwehr. (G. *breastwork*). Parapet.

Buheradas. (Sp.). A balcony similar to a machicolation used in Andalusian fortifications.

Bukeyashiki. (J.). A warrior's home in a Japanese castle.

Bulwark. (Middle English). Various forms of the word exist: *bulwerke, bullwork, bolwark, bulwarge, bulwarke, bullwark.*

(1) A fortification.

(2) A large defence work of earth or some other material.

(3) An earthwork in the form of a rampart, similar to a breastwork.

Bu man. (Ch. *cloth curtain*). A protective screen hanging from a pole mounted on the wall, made from multiple layers of cloth or netting.

See *'mu man'*.

Bunker. (Old Norse *bakki*, river bank). A shelter above or underground for people or troops to shelter from enemy fire and the weather, store food, weapons and supplies. Could be used as command centres with a telephone link to the rear for making reports and receiving orders. A bunker could be dug into soil or rock cut or built from concrete and reinforced steel.

See 'Anderson hut', 'flak towers (Germany)', 'flakturm', 'hochbunker', 'luftschutzraum', 'London deep-level shelters', 'Morrison shelter', 'Stanton air raid shelter', 'Winkel shelter'.

Burg. (*G. castle*).

(1) A German castle.

(2) A town or house with a fortified perimeter.

See *'burh'*, *'borg'*.

Burgen. See *'burgus'*.

Burggraben. (G.). Castle moat.

Burgi, Burgus.

(1) A Roman watch-tower, which was the forerunner of the *'bergfried'*, which was erected on the Germanic border of the Roman Empire.

(2) A walled suburb of a castle.

See *'burg'*, *'burgi'*, *'burh'*, *'limes'*, *'specula'*, *'vigilarium'*.

Burgward. A form of defended settlement dating from the 10th century CE located in the northeast of the Kingdom of Germany, consisting of a burg which defended a number of villages (wards).

Burh. (Old High German, burg). A fortified Anglo-Saxon town which was usually surrounded by a ditch and earthen rampart topped by a palisade. They were often built on former Roman or earlier fortifications, situated at the ends of estuaries or near fords and bridges crossing major waterways. Instigated by King Alfred the Great (871-899 CE) and carried on by his successors, were used to protect trade and culture from attacks, and as bases for launching assaults against Viking raiders.

See '*burg*'.

Burh-geat. The entrance tower or gate of an Anglo-Saxon fortified town or '*burh*'.

See '*burh*'.

Burj. See '*qalca*'.

Burster belt. A second layer of concrete, up to 1 metre (3 feet 3 inches) thick over a 1 metre (3 feet 3 inches) of sand over a concrete bunker, which was intended to detonate a shell before hitting a bunker. Dated to the 1900's CE, started being used by the French, later taken up by Belgium and Germany.

Buruj, Buruja. (Marathi). The part of an Indian fort including the towers, battlements and crenelations.

Butterfassturm. (G. *butter churn tower*). A defensive tower, where the upper section is smaller in diameter than the lower section. The ledge below the upper section is a '*chemin de ronde*' or fighting platform, while the upper section is useful as an observation tower. Such towers could be round or square in shape and appeared in the 14th century CE.

B-werk. (G.). A fortification of '*B*' grade of lesser strength.

Byoubuori. (J.). A Japanese castle wall set in a zig zag pattern providing multiple planes, increasing the amount of flanking fire which could be brought to be by the defenders' loopholes ('*aiyokoya-no-sama*') loosing arrows into the flanks of an enemy's attack. Also known as '*orihei*' (folding wall).

See '*aiyokoya-no-sama*', '*orihei*', '*sama*'.

C.

Cabulus. A 13th century CE term for a machine (possibly a trebuchet) that the soldiers of Philip Augustus used one against the walls of the inner wall of Chateau Gaillard.

Caer, Cair, Kair. (Welsh). A prefix denoting a citadel, fortress or stronghold.

Caislean. (Old Irish, *caisel*). An Irish castle dating from the 12th century CE which was based on the Norman '*motte and bailey*' castle.

Çakmak Line. (Turkish *Çakmak Hattı*). A defensive line constructed prior to the start of WWII by Turkey to deter or defend any attack by Germany on the Eastern Thrace border, running from the Sea of Marmara to the Black Sea, consisting of bunkers ,barbed wire and iron barriers based on the French '*Maginot Line*'.

See '*Anastasian wall*'.

Calotte. (Fr. *cap*). **See '*turret*'.**

Camp retranché. (Fr.).

(1) A French entrenched camp of the 18th century CE was large and included field works and semi-permanent works.

(2) In the 19th century CE this could mean a fortified city like Antwerp (Belgium).

(3) Whereas in the 20th century CE this was used as a '*base aéoterestre*' surrounded by earthworks.

Camouflage. (Fr. *to disguise, conceal*). Camouflage is to hide by providing: location and natural cover, artificial cover, camouflage paints and coatings, decoys, fieldworks, and smoke.

See '*camouflaging*', '*pillboxes*', '*spider hole*'.

Camouflaged tree. An observation post made from a metal cylinder made to look like a dead tree, used to observe the operations of the enemy during WWI, first used in 1916 CE. The camouflaging was based on an existing tree, which was cut down during the night and replaced with the camouflaged tree. Used by the armed forces of France, the United Kingdom, and Germany in trench warfare during World War I. Also known as '*dummy*' or '*fake trees*', '*false trees*', and '*observation trees*'.

See '*camouflage*'.

Camouflaging. Hiding the true nature of a thing. An example is England, fearing German invasion (1940 CE) built tens of thousands of pill boxes, often they were camouflaged as Beach Kiosks, summer houses (beach boxes), haystacks, piles of logs, parts of fairgrounds, or even parts of ruined castles (i.e. Pevensey Castle).

See *'camouflage'*.

Camouflet. (Fr. *kamufle*, camouflaged). A subterranean mine charge, or the cavity caused by the explosion of such a charge, used as a countermeasure against tunnelling or mining.

See 'mine', 'mining'.

Cang bing dong. (Ch. *troop or armament concealing cave*). A form of casemate.

Cannon. A large gun, using gunpowder to propel an object to cause damage and destruction, used in attack or defence. Also known as *'crakki'*, *'gonnes'*.

Cannon basket. *See 'gabion'.*

Canonnière. (Fr.). An embrasure in a transitional fortification in the late 15th century.

See 'gunport'.

Cantilevered pillbox. A British WWII pillbox used to defend airfields with a cantilevered roof with full circle embrasure, with a central cross shaped anti-ricochet wall, There are 53 examples that still exist.

See 'British hardened field defences', 'FW3', 'pillbox'.

Capannati. (It. *huts*). A casement that was not built into the walls of a work but placed across ditches and at the corners of strongholds. Free-standing *'capannati'* were also placed in ditches and on platform roofs.

Caponier.

(1) A covered passage that traverses a dry ditch from the curtain to the ravelin or from the covert way to an arrow or detached redoubt, covered on either side by a parapet and equipped with gun ports, through which cannon and muskets were used to sweep the ditch of enemy forces.

(2) A single parapet at the entrance of the ditch in front of a ravelin, equipped with a small cannon, which were used to dispute the passage of the ditch.

Cappanatto. (It.). A triangular bomb proof room provided with loopholes for defence, with an open rear to the interior of a fort.

Cardorna Line. ***See 'alpine wall', 'frontiera nord'.***

Carcass. A clay container hurled by mangonels which was filled with a mixture which would ignite on impact. Used by the Turks at the siege of Rhodes in 1480 CE.

Cardo maximus. (L.). The line which bisects the '*decumanus maximus*' or the '*centre line*' of a Roman fortification.

Carnot wall. A detached wall at the foot of a rampart, separated from the rampart by a '*chemin des rounds*' and provided with loopholes.

See '*Carnot*'.

Carnot. [Lazare Nicolas Carnot] (1753 – 1823 CE) Secretary of War to Napoleon Bonaparte, French Military Engineer, wrote "De la Défense des place fortes", 1810 CE.

See 'Carnot wall'.

Carrago. (L. *carrus, wagon*). A field fortification formed by surrounding an army with armoured wagons.

See *'gulaigorod', 'wagenburg'*.

Caravaserai. (Persian, kãrwãnsarãy) A fortified hostel along a trade route such as the '*Silk Road*'.

Casa-torre. (It. *towerhouse*). An Italian fortified house with a tower.

See 'towerhouse'.

Casatorri. (It. *family tower*). Italian slender towers, usually connected to the house by a drawbridge, with no ground floor entrance, which could be used as a refuge in times of strife. Like '*vepshi*' in Russia, and '*svaneti*' in Georgia.

See *'vepshi'*.

Casbah. ***See 'kasbah'.***

Cascan. The entrance to a gallery which provided a vent for an enemy mine.

Casemate, Casement, Cazemate, Cazematte.
(Fr. *casemate*, It. *casamatta*, prob. from casa house + matto, Fr. matta, mad, weak, feeble, dim. from the same source as E. -mate in checkmate).
(1) A chamber within a tower used to house artillery away from the elements such as catapults, Greek from the 4th century BCE.
(2) A gallery which was built at the base of a fortification's wall from which defenders could fire into the faces of surface miners and battering ram parties.
(3) A shaft having a number of underground branches which can be extended to intercept enemy mines.
(4) A magazine for storage of explosives.
(5) A place for quartering troops.

Casemate de Bourgres. A flanking casemate designed in the late 19th century CE.

See 'Atlantic Wall', 'festen', 'Maginot Line', 'traditore battery'.

Casemate type G. A small Dutch pillbox armed with a machine gun, dating to the late 1930's CE.

See 'casemate', 'cloche', 'pillbox'.

Casemated retrenchment. A retrenchment fitted with guns firing through embrasures from within vaulted casemates and placed behind the main line of fortifications.

Casemated tower. A large tower provided with numerous embrasures.

See 'artillery tower'.

Castrametation. The art of setting out a military camp or constructing a fortification traditionally has been called '*castrametation*' since the time of the Roman legions.

Cashel. An Irish defensive enclosure consisting of a dry-stone ring wall. An example: Cashel Murphy, Dingle, County Kerry, Republic of Ireland.

Cassaro. An Italian word meaning castle. Derived from the Arabic word '*al-Qasru*' for a castle.

Castella. A flanking tower or redoubt attached to the '*vallum*' of a '*castra stativa*', which was used to provide flanking fire at the gates, corners and at regular intervals along the ramparts, constructed of wood and earth or stone.

Castellae. A small Byzantine frontier castle.

Castellated.

(1) A building constructed in the form of a castle.

(2) A building provided with defences, such as battlements thus making it fortified.

(3) An area provided with a castle.

Castellated manor house. A fortified manor house.

See 'fortified manor'.

Castellation. A fortification, usually a castle.

Castelle. The Norman word for castle.

Castelletto. (It.). A small castle.

See 'castelletto', 'castlet', '*châtelet*', '*fortalice*'.

Castello.

(1) An Italian castle, which was mainly used as a fortified residence.

(2) A term which generally refers to an Italian fortified village.

Castellum. A small Roman detached fort or fortlet which served as a watch tower or signal station.

Castillos roqueros. (Sp. *rock castle*). A type of Spanish castle built on inaccessible sites on top of hills utilising the steep topography to the best advantage for defence, usually irregular in plan.

Castle. (L. *castrum*, fortress). A fortified building or set of fortified buildings used to provide both active and passive defence, as well as a residence for the castle's lord and household.

See 'fortification', 'keep', 'tower'.
See Appendix 1. What is a Castle?

Castlet. (E.). A small castle.

See 'castelletto', 'châtelet', 'fortalice'.

Castra.

(1) Roman *'marching camps'* were temporary structures built during military campaigns, with a ditch, rampart, walls and gateways. Also known as an *'auxiliary fort'*. Vindolanda was a Roman auxiliary fort or *'castrum'*, located just south of Hadrian's Wall in northern England (Hexham, UK).

(2) An Italian fortified settlement dating from the 10th to the 13th century CE. Usually sited on a hill top or spur, defences consisted of a surrounding curtain wall and a separate tower.

Castra aestiva. (L. *summer camp*). A Roman encampment used when camping in the warmer seasons.

See 'castra stativa'.

Castra hiberna. (L. *winter camp*). A Roman encampment used when camping in the colder seasons, provided with wooden barracks insulated with animal hides and straw.

See *'castra stativa'.*

Castra navalia, castra nautica. (L. *navy camp, nautical camp*). A Roman encampment for naval personnel.

See *'castra stativa'.*

Castra stativa. (L. *permanent camp*). When a Roman encampment was reinforced by strengthening the ramparts with a wooden stockade or stone wall, it became more than just a temporary defence works, it became what is known as a *'castra stativa'*.

The *'castra stativa'* served as a base from which actions could be launched into the surrounding area, and if necessary, a place to fall back to if things didn't go well. More elaborate *'castra stativae'* were provided with *'castellas'* or watch towers, which were used to flank the gates and at the corners and at regular intervals along the vallum.

Castron. A small fortified outpost of the Byzantine Empire. An example is Assos. located on the Aegean coast in the Ayvacık district of Çanakkale province, Turkey.

Cat. A mobile shelter for miners or ram. Also known as a *'sow'*, *'mice'* or *'weasel'*.

Catapult. A siege engine for throwing missiles.

See 'ballistra', 'trebuchet'.

Cat's ears. A form of caponier rarely used with two angled casemates projecting into a ditch, dating to the 1860's CE.

See 'caponier'.

Cavalier. (L. *caballus*, horse). A gun platform which is raised higher than the rest of the works, used to command the surrounding works, usually situated on a bastion or curtain but were also located in the gorges of bastions. Cavaliers were also built by besiegers to gain a commanding position.

See 'gun platform'.

Cavalier battery. A cannon battery in which the gun platform was raised above the level of the bastion.

See 'cavalier'.

Cavalier de trenchée. A temporary structure made of tiers of gabions to attack the defenders in the covered way and the place of arms, recommended by Vauban.

Cavalry forts. In 19th century CE the USA cavalry forts were used as bases for patrols of hostile territory and protection of wagon trains, consisting of a wooden stockade, typically measuring 183m X 114m (600 x 374 feet) , with at least one blockhouse in a corner, provided with stables, accommodation, offices, stores, magazine, band stand and flag pole.

Cave battery. ***See 'galleria cannonier'.***

Cave castle. ***See 'corona', 'höhlenburg'.***

Cawd. (Ar.). A field fortification made of kneeling baggage animals, such as camels.

See 'kuklos'.

Centre of the bastion.

(1) The intersection made by two demigorges.

(2) The point at which two adjacent curtains of a bastion intersect each other.

Cervi, Cervus. (L.). (L. *deer*, as in antler) The pointed timbers with forked tops used as '*chevaux-de-frise*' along the tops of the ramparts of the Roman fortification surrounding Vercingetorix during the siege of Alesia (52 BCE).

See *'cippi', 'chevaux-de-frise'.*

Cespites. The turves which were used to cover the outer face of the rampart of a Roman fortification, their purpose was to hold the ramparts in place.

See *'gazion'.*

Cha ban. (Ch. *slotting plank*). Acting like a portcullis, but more reinforced because they were also used as flood gates.

See 'orgue'.

Cha bei. (Ch. *raft stele*). **See *'cha ban'.***

Chain. Iron/steel chains to block access to harbours or waterways (i.e. Constantinople Golden Horn - estuary).

See 'boom defence'.

Chain tower. A tower used to house and defend a chain used to close off a harbour or waterway.

See 'boom defence', 'chain'.

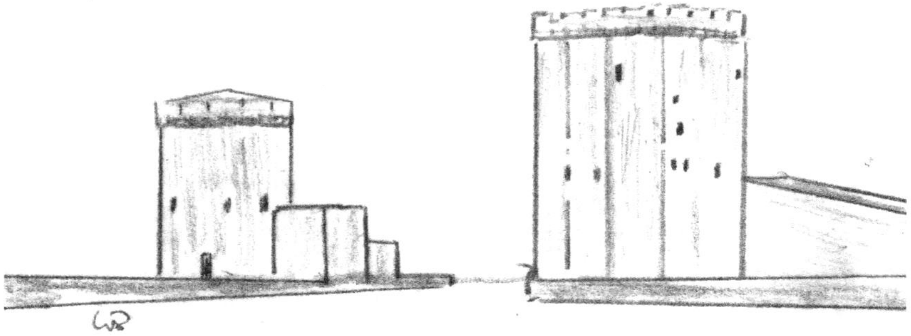

Chain tower, La Rochelle, France.

Chandelier. A wooden framed traverse built to protect besiegers while sapping.

See 'fascine', 'sapping' 'traverse (1)'.

Chamber gate. See '*kammertor*'.

Charika. (Sanskrit). A passage on the parapet for the movement of soldiers.

Chartaque. (Turkish, *chahartaq,* having four arches). A watchtower of the Ottoman Empire provided with palisading or other defence works such as '*abatis*', '*ramparts*' and '*ditches*'.

Chashi. A fortification dating from the late 6th to the 13th century CE, which consisted of a position surrounded by earthen ramparts and ditches. The ancestors of the Ainu, the '*Ezo*' built the chashi to protect their region and the population centres from the depredations carried out by the armies of the Yamato court. These fortifications are found on the Japanese islands of Hokkaido and the north east of Honshu.

Chateau. (O.Fr. *chastel*; L. *castrum*, fortress).

(1) Originally, a French castle, also known as a chateau-fort.

(2) A French country residence.

Chateau-fort. A fortified French chateau.

Châtelet. (Old French, *chastelet,* small castle).

(1) A small castle.

(2) A gatehouse or barbican.

See 'castelletto', 'castlet', *'fortalice'*.

Chatham forts. A series of polygonal detached forts using counterscarp galleries to control ditches built around Chatham, England, in the 1870s CE.

Chatris. (Indian). An Indian 14th century CE kiosk built over the main gate where the ruler could see and be seen.

Cheek. The side of an embrasure, which could be reinforced with *'fascines'*, *'gabions'*, *'sandbags'* or stonework.

See *'embrasure*'.

Chemin couvert. ***See 'covered way'.***

Chemin de ronde. A sentry path or a passage around the revetment of a rampart which was provided with a small parapet. The position was used by musketeers keeping an eye on the glacis so as to prevent the placement of scaling ladders by the enemy. Later, it was generally discarded because of its susceptibility to artillery fire.

Chemise. (L. *camisia*, shirt). The curtain wall, or the additional counterguard wall surrounding a keep or attached to it.

Chemise foot bridge. A footbridge connecting the curtain wall to the entrance of a Norman keep of the 11th and 12th centuries CE.

See 'flying bridge (2)'.

Cheng. (Ch. *fortified wall*). The Chinese defensive walls were built with foundations and walls of rammed earth, sometimes mixed with other materials from sand to rubble, and reinforced with wood piles, granite blocks and paved with brick. Blue bricks were used from the Tang Dynasty forwards to line the walls and bound with a rice based mortar. Most Chinese fortifications were built on the grand scale by the state (like the Great Wall of China) or providing a network of supportive fortresses, outposts, way stations and watch towers. Rarely are fortifications raised for the personal protection of an individual such as a member of the nobility.

See 'Han cheng', 'Hu cheng', 'lou cheng', 'man zhou cheng', 'yi cheng', 'yi zi cheng' 'zi cheng'.

Cheng lou. (Ch. *city tower*). A great tower sitting atop of the main gatehouse of a city, serving as command and observation platform, lacking in fortified structures and loopholes. Also known as a *'zhong men da lou'* (great tower of heavy gate).

See 'cheng'.

Cheng men. (Ch. *wall gate*). A gate of a gatehouse.

Cheng pu. (Ch. *wall shack*). A temporary or permanent shelter built on top of a city wall which is used for the shelter of a watchman. A permanent version on a bastion would serve as a watch tower or weapons storeroom. Also known as '*wo pu*'.

Cheng pu, on the city wall of Pingyao, China.

Cherta. Linear fortifications used by Tsarist Russia against the Tartar horseman, consisting of a ditch backed by palisades of pointed stakes, which linked stockaded towns, and natural barriers. Situated along the Ural and Tobol rivers then onto Kuznetsk in Siberia, 3000 km (2000 miles).

Chevaux de frise. (Fr. *Frisian horses*). A defensive obstacle which came in a number of forms, supplies with spikes used to quickly close a breach.

See 'frizzy horse', 'hersillion'.

Chikujõ-jutsu. (J.). The science of building and fortifying Japanese castles.

Chiritori. (J. *garbage*). A passageway in the earthen embankment or stone wall for the disposal of rubbish or hidden entrance to a castle, normally located inside the castle complex.

See '*chiriotoshi*'.

Chiriotoshi. (J.).

(1) A place built into an angled section of a castle wall, used to dump rubbish from a castle under siege, rather than the '*chiritori*'.

(2) A set of steps used covertly, leading to a dock constructed in the walls surrounding the central compound (*'honmaru'*). Also known as a *'fujounagashi'*.

See 'chiritori', 'honmaru'.

Cholli castle. A medieval Korean defensive wall built by the Koryo dynasty from the Yalu river to the east coast.

Chong yan. (Ch. *gun hole*). **See *'yan'*.**

Choujou. (J.). A lengthy defensive wall of earthen walls or stone ramparts, such as the *'Great Wall of China'* (J. *'Banri no choujou'*), another is the embankments built along the coast of Hakata Bay in defence against the Mongol invasions. Also known as *'chourui'* or *'bourui'*.

Chousenshiki yamajiro. (J. *Korean mountain castle*). A mountain castle constructed like a Korean one, consisting of earthen ramparts and stone work at the entrances, functioning like their Korean counterparts as fortified shelter of troops and civilians in times of strife. An example is Oonojou (Fukuoka Prefecture, Japan) with embankments over 9 kilometres long x 8 metres wide x 2 metres high (5 1/2 miles x 26 feet 29 inches x 6 feet 7 inches), dating to the 7th century CE.

See *'yamajiro'*.

Chuan lou. (Ch. *strung tower*). A semi-permanent terraced barracks on top of a city wall, taking advantage of the width of a wall top to provide hoarding, guard post and barracks on the wall.

Chui zhong ban. (Ch. *drooping bell plank*). Similar to *'xuan ban'*.

Church (*fortified*).

(1) A fortified church provided with defences such as battlements and arrow slits.

(2) A church which was used as part of, or as a fortification.

Ciglo. (It.) For glacis, after Francesco di Giorgione Martini (16th century CE).

Cippi. An obstacle consisting of sharpened branches set in a trench used to defend a Roman fort.

Circumvallate, Circumvallation (lines of). (L. *circum*, round; *vallum*, rampart).

(1) A siege works constructed by a besieging army to protect their camp from sorties by the besieged garrison, consisting of encircling earthen ramparts and entrenchments. Attacks against the besieged position could be launched from the line of circumvallation and further advancement of the earthworks could be constructed. In conjunction with the line of contravallation, these fortifications enable the besieging

army to stop the flow of information, supplies and reinforcements reaching the besieged.
(2) The process of carrying out the construction of lines of circumvallation.

See 'circumvallatio', 'contravallation', 'counter vallation'.

Circumvallatio. The ditch and rampart which was constructed by a besieging Roman army, which encircled a besieged city fortress, situated between the city and the Roman positions.

See 'circumvallate'.

Cistern. An open or covered tank for collecting and storing water. Examples are the Crusader castles, Masada (Israel) and Constantinople (Istanbul, Turkey).

See *'berquil'*.

Citadel. (L. *civis*, citizen). A fortress in or near a city which was used to control the city and its inhabitants; providing a strong defensive position, and once the outer defences had fallen it could be used as a final refuge.

Clausurae. (L. *closing*). A linear fortification in stone or earth which could be provided with gates and towers, used to create a barrier across an access point (such as a valley) to defend a frontier of the Roman Empire. Two examples are: 1) Clausura Bir Oum Ali, which was part of Limes Numidiae, Tunisia; and 2) Claustra Alpium Iuliarum (L. 'Barrier of the Julian Alps'), Slovenia. Also known as a '*cross wall*'.

Clavicula. (L. *small key*). The external or internal right-angled extension of the rampart at the gateway of a Roman fort, which made the assailants turn to their left thus exposing their unshielded right side to the fire of the defenders on the ramparts and the '*castella*'.

See 'bent entrance'.

Cliff castle, Cliff fort. A fort which was situated on a spit of high ground and was protected on three sides by sheer cliffs, leaving only one line of approach which was defended by a rampart and ditch across the narrowest neck of the peninsula. If suitable headlands were not available, straight lines of cliff edges could be used by raising '*L*' or '*V*' shaped ramparts. Also known as a promontory fort, the majority occur on Britain's western seaboard. Some can also be found on heights above rivers further inland. They were easily constructed and could be defended by only a small force, however, they had one major flaw, if an enemy did manage to break in it would have been impossible to escape.

Cloche. (Fr. *bell*).

(1) A fixed armoured cupola or dome of the French *'Maginot line'*, embedded into a concrete block.

AM (*'armes mixte'*) cloche. A cloche mounted with a 25mm anti-tank gun and paired machine guns in two separate ports.

JM (*'jumelage de Mitrailleuses'* (twin machine guns)) cloche. A non-retractable, and a non-rotating cupola of steel alloy provided with heavy twin machine guns.

LG (*'lance-grenade'*) cloche. A cloche with a grenade launcher, used for close defence.

VDP (*'vision directe périscopique'*) cloche. A cloche provided with three embrasures for direct vision.

VP (*'vision périscopique'*) cloche. A cloche that lacked direct observation ports, which had a lower profile than the *'VDP cloche'*.

(2) **Czech cloche**. A cloche provided with defensive weapons or observation.

See 'cloche'.

(3) **German cloche**. A cloche with a fixed turret.

See 'beobachtungsglocken', 'cloche'.

Cloche Guet-Fusil-Mitrailleur (Cloche GFM). (Fr. Watch-Rifle-Machine Gun Bell). A type of small non rotating cloche for observation and defence with a machine gun on the Maginot Line.

See 'cloche', 'Maginot Line'.

Cloche Parmart. A French cloche dating from WWII, designed by Léon Pamart, armed with one or two machine guns, consisting of a metal cupola with one or two portholes, concrete screens, periscope and venting. An example can be found at Fort Souville, near Verdun, France.

See 'cloche'.

Clover leaf keep. *See 'device forts', 'quatrefoil keep'.*

Coastal battery. A type of gun emplacement used to control coastal shipping and possible landing sites such harbours, river mouths and anchorages.

Coastal fort. A fort used to control shipping and defend landing sites, harbours, navy bases, provided with gun emplacements, powder magazines, storage, accommodation for troops, loop holes for close range defence, and landward side was often defended by ditches, covered way and glacis.

Coastal fortification. A fortification built on and to defend the coast.

See 'Atlantic wall', 'beacon', 'choujou', 'device forts', 'coastal battery', 'coastal fort', 'concrete battleship','emergency coastal batteries', 'fire control tower', 'küver', 'martello towers', 'pile barrage', 'saxon shore forts', 'sea fort', 'Skäne line', 'torre saracinesche', 'tour-modèle'.

Coercion castle. See '*zwingburg*'.

Coffer. (Old French, *coffer, chest*).

(1) A trench in the bottom of a dry ditch which was used for drainage.

See 'batardeau', 'cunette'.

(2) A work similar to a caponier which traverses a dry ditch, provided with a parapet with embrasures, used by the besieged to dispute the passage of the ditch by the enemy.

See 'caponier'.

Coffer dam. A fortified dam retaining the water of a moat.

Coffre. (Fr.). A counterscarp casement to defend the ditch or exterior of a fortification.

See 'casemate', 'counter scarp'.

Cohorn (Menno van Coehoorn) (1641 – 1704 CE). A Dutch military engineer, a rival of '*Vauban*', invented a portable siege mortar, and served the Duke of Marlborough in the War of the Spanish Succession.

Column of fire. The amount of fire by the soldiers entrenched behind the parapet directly to their front when under attack. By changing the '*faces*' and '*flanks*' of the parapet each segment of parapet produced their own '*column of fire*', and thus produced interlocked fire from adjacent segments of the parapet.

See'arc of fire', 'parapet'.

Coles turret. An armoured revolving turret with a wide arc of fire, first patented by Captain Coles of the British Navy in 1859 CE, providing good protection for the ammunition, crew and gun, and improving naval and fortress armament.

See 'turrent'.

Combustibles. Combustible materials used in missiles, used to fire mines and other structures. An example was the Hungarians attacking Pavia in Italy in 924 CE, by throwing lit torches over the walls setting fire to the city inside the walls.

Command. When any work is constructed higher than another the work overlooking the other is said to command it.

See 'datum'.

Commanding position. A position that has a view of the surrounding country and enables an army to provide '*plunging fire*'.

Commission d'Organisation des Régions Fortifiés (CORF). The French commission was set up in 1927 CE to prepare and plan for the Maginot Line.

See 'Maginot Line'.

Communication. A passage which runs between works, generally covered by a parapet on either side, providing for the safer movement of troops.

See 'caponier'.

Communication Trench. The trench used by troops to move between the front and rear trenches, also used to move ammunition, supplies and wounded soldiers. Also known as '*saps*'.

See 'sap', 'trench', 'zig zag'.

Concentric castle. The castle type was introduced into Europe in the late 12th century CE which consisted of two or more complete rings of bastioned curtain wall within one another, each increasing in height towards the keep in the centre. A variation on this design was where the gatehouse doubled as the keep.

The aim of the design was to present the enemy with sides of equal strength thus allowing a faster response to attack by the garrison. With each ring of defence increasing in height towards the centre the defenders on the higher walls and towers could fire at the enemy over the lower defences, so that the amount of missile fire was increased compared to contemporary designs.

If the outer defences fell to the enemy, they would be confronted by yet another line of defence; and each set of walls had to be taken completely before the attackers could move on to the next, thus increasing the losses to the besieging force as the siege continued. Also known as a '*multiple castle*', Caerphilly Castle in South Wales is a good example.

Caerphilly castle, Wales, UK.

See '*defence in depth*', '*lines of defence*','quadriburgia'.

Concertainer. A wire mesh gabion with fabric liner, filled with rocks and dirt and used as part of a temporary defensive wall or enclosure during the war in Iraq and Afghanistan.

See '*blast wall*', '*hesco barrier or bastion*'.

Concertina wire. Coils of barbed or razor wire which when unravelled are similar to a concertina, used in construction with plain barbed wire and steel pickets to form wire obstacles and entanglements. Also known as *'dannert wire'*.

See *'barbed wire'*. *'entanglement'*.

Conclave flank. The inward curve of the flank of a bastion (i.e. Naaden in the Netherlands).

Concrete. Building material consisting of water, cement (lime) and sand or gravel used by the Romans to the present day. which can be used to make a range of structures from pill boxes to flak towers. First example of use of concrete as a revetment material for ditches in a British fort, 1864 CE at New Haven Fort, Sussex, UK.

Concrete battleship. A coastal fort in the shape of a battleship, provided with armoured turrets (i.e. Fort Drum, Philippines).

See 'sea fort'.

Console. ***See 'corbel'.***

Constantian Wall. A defence work started by Constantine, consisting of a curtain wall with rectangular towers, dating to 324 CE thus pre-dating the Theodosian land walls. The approximate route ran from the Church of St. Anthony at the Golden Horn to between the later sea gates of St. Aemilianus and Psamathos.

See 'Herclian Wall', 'Theodosian Land walls'.

Continuous line. A set of fortifications presenting a continuous front to the enemy, be it parapets or curtains connecting the *'artillery strong points'*.

See *'lines of intervals'*.

Continuous wall walk. An uninterrupted wall-walk around a curtain or tower wall, allowing the fast movement of equipment and troops to any quarter under attack.

See 'bastille'.

Contour fort. A hill fort which was constructed by reinforcing the natural defences of the position by digging a bank and ditch along the contour line surrounding a hill.

See 'hill fort' (Maiden castle, UK).

Contravallatio. A series of fieldwork consisting of earthen ramparts and ditches excavated parallel to the *'circumvallatio'* but facing outwards, which were used to protect the Roman encampments against the attack of a relieving army; and in conjunction with the *'circumvallatio'*, prevented supplies and reinforcements reaching the besieged city.

See 'circumvallate', 'circumvallatio', 'contravallation'.

Contravallation (*Lines of*). A chain of redoubts and breastworks constructed by a besieging army which encircles a fortification, situated parallel to the circumvallation but facing the opposite direction. This set of works was used to protect the besieger's camp and positions from assault by a relieving army.

See 'circumvallate', 'counter vallation'.

Contre approaches (counter approaches). Tasks that consist of the besieged trying to move forward from the defence works on the soldiers who wanted to block them (the besiegers).

Contre escarpe (counter scarp). The outer slope or wall of the ditch or moat of a fortification.

Contre garde (counter guard). A counter-guard is in bastioned fortification an external work of protection of a bastion or a half-moon in the shape of an inverted '*V*'. It is an evolution of the face cover.

Cop. A term used to refer to a merlon.

See *'merlon'*.

Corbeil, Corbeille. (Fr. *basket*). A small wicker basket made wider at the top, similar to a *'gabion'*, filled with earth and placed along the parapet, the tapered shape allowed for musketry loopholes between the corbeils.

See *'gabion'*.

Corbel. (L. *corvus*, crow). A projecting structure made of stone of wood located near the top of a wall, which were used to support such structures as; breteches, hoardings and machicolations.

See 'bracket'.

Corbelled bartizan. A bartizan built out on corbels, which was used to provide flanking fire thus reducing dead ground.

See *'bartisan'*.

Cordon, Cordone.
(1) A row of stones along the line of a rampart.
(2) A continuous rounded coping at the top of a masonry revetment.
(3) A chain of military posts.
Corduroy road. A roadway constructed of logs laid parallel inside or outside a fortification, used on swampy ground or in wet weather.
Cormontaigne, Louis de. (1695 – 1752 CE). A French engineer involved with various sieges, later became director of fortification for towns like Moselle, France.
Corner buttress turret. A corner turret projecting only a small distance beyond the walls of a Norman rectangular keep; the turret had little if any military value as far as flanking ability was concerned.

See *'square keep'*.

Corner flanker. A flanking tower projecting from the corner of a keep, which was used to flank the wall between the towers, and to protect the vulnerable corners from sapping by miners and battering rams.

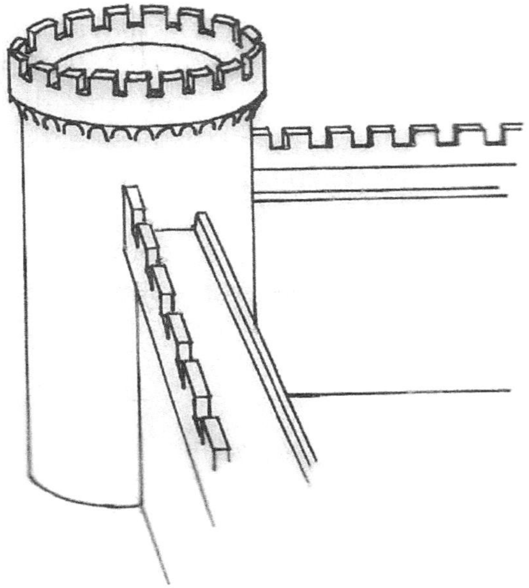

Cornichon. (Fr. *pickle*). A French hornwork with two short wings (Fr. ouvage à corne, horn work).

See 'hornwork'.

Corona. (It. *crown*). A type of castle built into a cave, such as San Gottardo in Mezzocorona, Italy.

See 'höhlenburg'.

Corps de garde. (Fr. *guardhouse*). A post for a guard of a citadel or fort.

Corradino Lines. (Maltese, *Is-Swar ta' Kordin*). A line of fortification built by the British on the Corradino Heights in Paola, Malta, consisting of a '*V*' shaped polygonal-style trace surrounded by a ditch, dating from 1871 to 1880 CE.

See '*Cottonera lines*', '*Floriana Lines*'.

Corridor. A covered way.

See '*covered way*'.

Cortina. (It. *curtain*). The main wall of a bastioned work, normally between two bastions.

Cottonera Lines. (Maltese, *Is-Swar tal-Kottonera*). A bastioned enceinte consisting of eight bastions, two demi-bastions and nine curtain walls built in the 17th and 18th century CE on high ground in Bormla and Birgu, about 5km (3 miles) in length. Also known as '*Valperga Lines*' (Maltese: *Is-Swar ta' Valperga*).

See '*Corradino lines*', '*Floriana Lines*'.

Couillard. A smaller version of a counterweight trebuchet with a single frame instead of the usual double '*A*' frames. The counterweight is split into two halves to avoid hitting the centre frame.

See '*trebuchet*'.

Counter approach. A temporary work usually consisting of entrenchments built by the besieged forward of permanent fortifications, to check the approach of an enemy army.

Counter arch. An arch connecting two counterforts.

See '*counterfort*'.

Counter castle. As part of long term sieges, counter castles were raised to protectively house besieging soldiers, which were so placed to block access roads, to stop resupply of food and soldiers. Examples were erected by William Rufus at Bamburgh (1095 CE), and Henry 1 at Arundel in England (1102 CE). The remains of a counter castle at Bentley in Hampshire (UK) have been excavated and appear to be a '*Motte and bailey castle*' in design (1147 CE). See Stamper (1984 CE).

See '*malvision*'.

Counterfire room. See '*counterscarp gallery*'.

Counterfort. An inner buttress used to strengthen the masonry of a rampart's revetment.

Counterguard. An outwork, triangular in shape, open at the rear, situated before the face of a '*bastion*', '*ravelin*' or '*redan*', used to protect these works from enemy cannon fire.

Counter mine. A mine dug to intercept the mine of a besieging force, so that a counter attack could be launched, the objective of which was to collapse the mine of the besiegers.

Countermure. (Fr. *contremur*).

(1) A strengthening wall built in front of another wall.

(2) A wall which was built behind that of another as a reserve defence.

Counterscarp. The outer side of a ditch of a fortification, in some permanent fortifications it was faced with stone to make entering and retreating from the ditch more hazardous.

Counterscarp bonnet. A bonnet used to defend the ditch, constructed in the counterscarp wall salients, and is connected to the counterscarp gallery by covered wall walks.

See 'bonnet'.

Counterscarp gallery. A gallery which is located in the counterscarp of the salients, used to flank the ditch of a fortification.

Counter salient. A salient in the opposite direction.

Counter trench. A trench which was excavated by the besieged in opposition to those of the besiegers. The parapet was turned against the enemy's approaches, and was flanked by artillery providing enfilade fire so that if the enemy took the trench it would be rendered useless.

See *'trench'.*

Counter vallation, Contravallation (*Lines of*). The lines of earthworks thrown up by besiegers to protect their camp and positions from attack from a relieving army, used in conjunction with the lines of circumvallation during a siege of a fortification.

See *'circumvallation', 'contravallation'.*

Counter work. Any type of work which was used to oppose those of an enemy; an opposing work.

Coupure. (Fr. *couper*, to cut).

(1) An entrenchment made by the besieged behind a breach in the defences.

See *'entrenchment'.*

(2) A passage through the glacis to allow the defenders to sally forth to attack an approaching enemy force.

Coursiéries. Mobile wooden parapets attached to the inside walls, which were used in France around the 11th century CE.

Court. A space enclosed by walls or buildings.

See *'bail', 'bailey', 'barbkin'.*

Courtine. (Fr. *curtain*). A section of wall between two towers or bastions.

Courtine brisée. (Fr. *broken curtain*). A section of wall that is angled to improve flanking fire.

Courtyard castle. A castle where the domestic buildings were integrated into the curtain walls and arranged around a central courtyard.

See'courtyard castle' 'keepless castle', 'quadrangular castle', 'quadrilateral castle'.

Couvreface. (Fr. *face cover*). A fortification consisting of a low rampart with breast work and ditch, in front, which was built in front of a *'fortress', 'bastion'* or *'ravelin'*. Used in Dutch and French fortifications from the 17th to the 19th centuries CE, only suited for infantry with muskets.

Couvre porte. (Fr. *door cover*). A work used to cover the approach to the main fortress.

See 'covered way'.

Covered way. A pathway or road running along the top of the counterscarp, provided with a protective embankment which formed the crest of the glacis. The embankment gave the soldiers standing in the covered way some protection from enemy fire, being high enough to obstruct the besiegers view.

See 'Covert way'.

Coverport. A small defensive work placed in front of a gate.

Cover trenches. Trenches used to provide protection to the support and reserve troops, and were not designed to be fired from.

See *'trench'.*

Covert way. A pathway in front of the ditch protected by a parapet which formed the crest of the glacis. The covert way was used by soldiers whose job it was to break up the general line of assault.

See 'covered way'.

Crakkis of War. Early reference to cannons used by English in Scotland during the reign of Edward III.

See *'cannon'.*

Crannog. (Ir. *crann*, tree). An Irish and Scottish fort or lake dwelling dating from the Bronze Age and used up to the 16th century CE which was located on an islet, often an artificial one which was built up in a lake or marsh. Also found in Denmark.

Cremaille. (Fr. *pot hook*).
(1) A zig zag line of a fortification.
(2) A sawtooth pattern on the inside line of a parapet.

See '*dentate*'.

Crenaux. A small loophole in the wall of a fortification, constructed with a widening aperture on the inside to allow the defenders a larger field of fire, but also allowing for the protection of the defenders.

See 'loop'.

Créneau de pied. (Fr. *foot niche*). A type of machicoulis dated from the 19 and 20th centuries CE, used to fire down at an enemy below or drop grenades down.

See 'goulotte à grenade'.

Crenel, Crenelle. (L. *crena*, a notch).
The part of a parapet which is indented alternating with the solid uprights called merlons, which allowed the defenders to fire at the enemy while gaining protection from the merlons against the returned fire.

See '*battlement*', '*embrasure*', '*merlon*', '*reveal*'.

Crenulate. To fortify or embattle.
Crenelated. Fortified or provided with crenels.

See 'battlemented', 'embattled'.

Crenelated gallery. See '*reverse fire gallery*'.

Crenelation. A parapet consisting of merlons and crenels. The '*Licence to crenelate*' was a royal licence giving permission to the holder to build a fortification or to fortify a present building.

See 'crenel', 'merlon', 'parapet'.

Crenel shutter. A wooden shutter which covered a crenel and was used to defend a crenel, hinged at the top, which enabled the defender to open the shutter to fire at the enemy, while gaining protection from the shutter in the closed position. An example is Gravensteen castle in the city of Ghent, East Flanders in Belgium, dating from 1180 CE.

See 'crenel'.

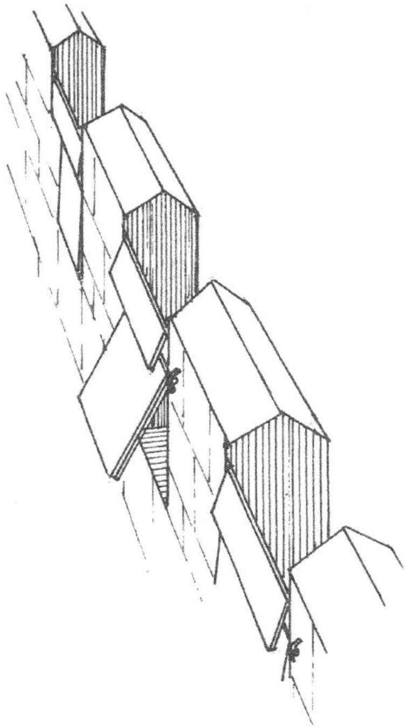

Crest. The apex of the glacis, either formed by the parapet of the covered way, or where the glacis meets the top of the counterscarp. Also known as a '*ridge*'.

Crib work. A supportive construction system for fortification walls, using wooden beams, embedded, connected and arranged within a masonry wall, increasing the plasticity to the wall, improving survivability against earthquakes.

See '*intra-mural timbers*'.

Crimean defence line. Defences built by the Russians in 2022 and 2023 CE to defend Crimea along the beaches of the Black and Azov seas, and along the border with the Kherson Oblast, from attack, including trenches, anti-tank bunds and artillery emplacements.

See 'linear fortifications'.

Crochets. Passages round the head of traverses to allow access to the covered way.

Cromwellian Forts. Used to pacify Scotland and Ireland, late 16th and 17th century CE, consisting of isolated forts, rectangular or square in shape, with bastions at each angle. An example is Greenfort at Sligo (Ireland).

Cross and orb gun loop. A gun loop provided with a cruciform sighting slit above the circular aperture for the cannon. The design came from adapting the cruciform arrow slit shape for the sighting slit, it provided effective sighting and a vent for the gases produced by cannon fire.

See 'gun loop', 'gun port'.

Crosslet slit.

(1) An opening in the wall of a castle in the form of a cross for firing crossbows at an enemy without.

(2) A cruciform aperture in the wall of a castle which was designed for the use of both bows and crossbows. Also known as a *'crosslet arrow loop'*.

See *'balistraria', 'loop'.*

Cross slitted gun port. ***See 'cross and orb gun loop', 'gun port'.***

Cross wall.

(1) A strong wall which divided a keep in half, the wall provided additional strength to the structure.

(2) The wards between the walls of a type of concentric fortification were sectioned off by cross walls which ran between the outer and inner walls. If the enemy broke through the outer defences they were more easily disposed of (in a killing ground).

Examples can be found at:

- Tiryns, Built for King of Argos , 13th century BCE;
- Sinjerli, Nth Syma, dated 7th Century BCE;
- Sahyun castle, Syria, 12th century CE;
- Acre, added by Richard Coeur de Lion, 12th century CE.

(3) A linear fortification closing off access to an area.

See 'clausurae'.

Crotchet. A pathway around a traverse in the covered way, normally only wide enough for only one soldier to pass abreast.

Crow. A hook on a giant long pole used to catch up men on the battlements. Used at Lundow in the UK in 1139 CE. The tale that Orderic Vitalis records of Walter FitzWilliam was captured from the battlements by a crow in the 11th century CE. Henry, the son of the king of the Scots was dragged from his horse by a *'crow'* at Ludlow in 1139 CE, but managed to escape.

Crown. To construct a work on the apex of the glacis or the top of a breach.

Crownwork. An outwork consisting of two long flanks and a front with a bastion and two demibastions, thus forming the shape of a crown.

Cruciform arrow slit. An arrow slit provided with a horizontal slit, which was used to increase the available field of fire, dating from the end of the 12th century CE.

See 'arrow loop', 'arrow slit'.

Crusader castle. A castle erected or modified by the *'Crusaders'* to occupy and control their territorial gains in the Holy Land. An example is Krak des Chevaliers in Syria.

Cunette. (It. *lacuna*, ditch). A trench or drainage ditch sunk into the floor of a main ditch.

See *'coffer'*.

Cupola. (It. *dome*). General term for a domed, rotating, armoured gun mounting; may also be used to mount searchlights or as an observation point.

Curtain. (L. cortina, Fr. courtine). ***See 'curtain wall'.***

Curtain angle. The angle made by a flank (a work such as a bastion) with a curtain.

Curtain tower. A tower which was a part of a curtain wall, from which defenders could provide flanking fire to the curtain wall since they usually projected further into the field.

See 'bastion', 'mural tower', 'wall tower'.

Curtain wall.

(1) A fortified wall which enclosed a bailey or ward.

(2) The fortified wall which ran between two towers.

(3) The rampart which ran between two bastions.

See 'chemise', 'rampart'.

Curtes, Curtis. A Frankish fortification of the 8th century CE constructed on a rectangular ground plan, using ditches and earthen ramparts topped by palisades.

Curtes regia. A fortified residence of a Carolingian king.

Curticula. A forward element of a Frankish fortification.

Curved merlon. The design of merlons underwent a change due to the fact that they were one of the most vulnerable parts of a fortification to artillery fire. The merlons were thickened and curved backwards, and the actual number of crenels was reduced. Also known as shot deflecting battlements.

See 'merlon', 'shot deflecting battlement'.

Curvilinear wall. A wall consisting of round bastions which were built so close together that only a small portion of each protruded. An example is Chateau Gaillard, Normandy, France.

Cuslett. Pickets at the bottom of a dry ditch or trench.

Cut. A trench across a terreplein to stop attacking troops advancing from a breach.

See *'coupure', 'cutting', 'retirade'.*

Cutting. A narrow cut through the parapet of the *'covered way'* or the crest of the *'glacis'* to allow troops to quickly launch counterattacks. Also found through the parapet of the *'place of arms'*.

Cuvette. *See 'cunette'.*

Czech Hedgehog, Hedgehog. Static anti-tank obstacle, first used on the Czech defensive line (1935 – 1938 CE), first made in concrete, then made of three bars of iron welded together, two bars in a cross with a third bar attached at right angles to the first two bars, also adopted by Russians and Germans in WWII.

See 'anti-tank obstacles'.

D.

Dacoit fence. A type of fencing on British 19th century CE forts, consisting of spiked steel bars on a frame, anchored into the ground by concrete, used to defend military sites as a part of a '*Twydall profile*', usually on the outside of the glacis. There appears to be a piece of dacoit fence remaining extant at the site of the Home Chain Radar Station, RAF Bawdsey, UK.

See 'fence', 'twydall profile'.

Daidokoro yagura. (J.). A kitchen turret.

See '*yagura*'.

Dai'in-no-koguchi. (J.). An early type of entrance to a castle, which consisted of a passageway through an earthen embankment, lacked protective features and was improved with cross over embankments and barriers.

See '*koguchi*', '*kuichigai*', '*shitomi*', '*shouheki*'.

Da jiang jun chong yan. (Ch. *great general gun hole*). A gun hole used to accommodate a '*da jiang jun pao*' or a '*Fo lang ji chong yan* (Frankish engine gun loop), which was a medium sized gun loop for a breech loading swivel gun.

See '*yan*'.

Dame. A solid turret forming an obstacle on the ridge of a '*cofferdam*', used to block the path of besiegers.

See 'batardeau', 'cofferdam', 'monk'.

Danevirke, Danework, Dannevirke, Danavirki, Danewerk. (Danish, *Work of the Danes*). A system of earthwork ramparts and ditches against Frankish attacks, built in three phases between 737 and 968 CE, running 30 km (19 miles) from Hedeby (former Viking trading centre) near Schleswig to the coast of the Baltic sea, in Germany.

Danske. The sewage tower of a castle constructed by the Teutonic Knights, which was detached from the main body of the castle, and was only accessible by a gallery supported by five large arches.

Dashibei. (J.). A wall projecting outwards from the top of a castle's embankment ('*inubashiri*'), which enables defenders to observe and attack the enemy below.

Datum. The natural level of the site, from which '*command*' is calculated.

See 'command'.

Davit. A type of jib crane used to raise ammunition from a store to artillery in a fortification.

Dead angle. The angle between the top of a fortification wall and where the defenders on the top of the wall ceased to be able to bring their fire to bear upon the enemy approaching the walls.

See 'dead ground'.

Dead end corridor. A corridor which was situated on the inside of a fortifications gateway, which ended in dead ends, thus when an enemy force broke through the main gate and entered the corridor they were confused and trapped in a specialized killing ground.

Dead ground. Any area of a fortification where the enemy could not be attacked by the defenders, which was due to either the design or the siting of the work. Dead ground could be reduced by the use of hoardings, flanking towers or some other flanking defences.

See 'dead angle', 'flanking'.

Deblai. The ditch which was excavated when constructing the earthen ramparts of a work, the earth from the ditch was thrown inwards to form the rampart.

See 'remblai'.

Decauville railway. A narrow gauge railway used for logistics by the French military expeditions to Madagascar and Morocco, and later by the French and British in WWI, utilising readymade and transportable sections of track, designed by Paul Decauville.

Declivity. The open slope from the covered way to the country surrounding a fortification, on which an assaulting force was completely exposed to the fire of the defenders.

See 'glacis'.

Decumanus maximus. The centre line of a Roman fort.

Defence, Defense. The fortifications that provide cover for the opposite flanks, parapet, casements, etc.

Defence in depth. A fortification is said to have '*defence in depth*' with multiple lines of works, such as concentric castle or '*multivallate*' hill fort.

See *'concentric castle', 'hillfort', 'lines of defence', 'Theodosian land walls'.*

Defence (Lines of). A series of fortified outposts.

See 'lines of defence (3)'.

Defensible barracks. Casemated barracks came in a variety of shapes and sizes, used in the 1800's CE to provide safe and defensible accommodation for soldiers of a fortification.

Defensive fighting position. A type of earthwork constructed for defence, generally large enough to accommodate anything from one soldier to a fire team.

See *'fighting pit', 'foxhole', 'gun pit', 'rifle pit', 'slit trench', 'tobruk'.*

Defensive wastes One way of defending territory or a castle was to implement *'defensive wastes'* or what is also known as a 'scorched *earth policy*', and can involve the natural and built environment, including: bringing the livestock within the walls, or slaughtering animals and burning other foods stuffs; bring in the harvest, then burning hay and other fodder, and cutting down fruit trees and vines; seeding /ploughing fields with salt; knocking down buildings and cutting down trees and shrubs that would provide cover or shelter; blocking access to or poisoning/fouling water sources. Also known as a *'scorched earth policy'*.

See *'no man's land'*.

Defencework. Literally, a fortification; a work which was used to defend a position.

See 'works'.

Defilade. To secure a work or part of it against enfilade fire.

See 'enfilade', 'traverse (1)'.

Deguruwa. (J.). A castle compound that sticks out from one of the main compounds and is used to provide protection with a flanking effect.

See *'demaru', 'honmaru', 'kuruwa', 'ni no maru'*.

Dehor. (Old Fr. *outside*). Any work (e.g. crownwork, hornwork, halfmoon, ravelin, etc.) which was used to increase the defensibility of a fortification.

Dejiro. (J.).

(1) An outer citadel of a Japanese castle.

(2) A Japanese castle built on the borders or on a strategic site in a domain which would protect the main castle of a domain.

(3) A small Japanese castle or tower project from a larger castle. Also known as a *'demaru'*.

See *'demaru', 'shijou'*.

Deliberate entrenchments. Defence works built in anticipation of an attack on a town, depot or bridge, usually with a front ditch. Works could be improved if required for a longer period, including magazines, bomb proof shelters, sods added to parapets, revetments strengthened with planks of wood or stone, wells dug, roads and ramps paved, drainage put in place, barracks and storehouses built.

See ***'permanent fortification'***, ***'temporary trenches'***.

Deliberate fieldwork. Temporary field works and entrenchments could be improved with the addition of detached works for artillery.

See ***'permanent fortification'***, ***'siegeworks'*** ***'temporary trenches'***.

Demaru. (J.). A small tower or castle attached or projecting from a larger castle.

See ***'dejiro'***.

Demi-bastion. A half bastion consisting of one face and one flank.

See ***'bastion'***.

Demi-caponier. A caponier which was only protected on one side.

See ***'caponier'***.

Demi-gorge.

(1) A half a gorge; the area between the curtain angle and the centre of the bastion.

See 'gorge'.

(2) The rear entrance of a bastion.

Demilitarized zones. (DMZ or DZ). An area designated by agreements or treaties between countries or opposing groups bar the building of military installations, activities of a military nature or any military personal. A DMZ may be an established frontier between two countries such as North and South Korea forming the Korean Demilitarized Zone.

Demi-lune. (L. *luna*, moon). A work in the shape of a halfmoon which was used to defend the entrance of a fortification. Later, the demilune developed into a detached work called a ravelin which was situated within the line of the main ditch and was formed by two faces meeting in an outward angle, its purpose was mainly to cover the curtain it fronted and to prevent the flanks from being attacked from the side. Also known as a '*mezzaluna*' or '*halfmoon*'.

Demining. Removal or purposeful detonation of mines.

Demi-revetment. The partial facing of a rampart with concrete or masonry when it was raised only as the cover in front.

See 'revetment'.

Demolition passage. An underground work built with the defensive perimeter (i.e. below the tenaille lines, counterguards and ravelins) which can be demolished to render useless if overrun by the enemy.

Deng cheng ma dao. (Ch. *wall scaling horse route*). A stairway or ramp which can be used by cavalry to access the wall top. Deng Cheng Ma Dao can be shortened to '*Ma Dao*' (Ch. *horse route*). Also known as '*Shang cheng deng*' (Ch. *wall accessing stairway*).

Dentate. A zig zag or saw toothed parapet designed to the left or right across the front of the work, designed to create interlocking fields of fire.

See '*cremailliere*'.

Denti di drago. *See 'dragon's teeth'.*

Denti di drago near Graum im Vinschga, Italy.

Descents.

(1) The cavities and mines made by members of a besieging force, which undermined the ground under the walls of the fortification being besieged.

(2) Saps excavated in the counterscarp beneath the covered way, made by the besiegers to enable them to cross the ditch.

Detached bastion.

(1) A bastion isolated from the main works by a ditch.

(2) Counterguards provided with flanks are also called detached bastions.

See '*bastion*'.

Detached redout. A small work situated beyond the glacis but within musket shot of the covert way, similar to a ravelin with flanks.

See 'redout'.

Detached scarp. A wall in a ditch which was separated from the glacis of the parapet by a pathway.

Detached work. An outwork which was separated from the main works, such as a *'ravelin'*.

See 'forework', 'outwork', 'ravelin', 'works'.

Deterrence effect. The effect of strong appearing military architecture which psychologically deters an enemy from attaching, especially used on gateways.

Detinets. (Russian). A fortress. **See *'kremlin'*.**

***Device* Fortification.** A document was produced in 1539 CE called the *'Device of the King'* (Henry VIII) regarding advice on improvements and to search out positions of coastal defences of England and South Wales, resulting buildings to deny an enemy landing sites and harbours/anchorages in support of an invasion in the construction of squat artillery forts with circular bastions, including; Deal, Walmar, Sandown, Portland, Camber, Sandgate, Pendennis, Hurst, St. Mawes and Calshot.

Dezama. (J.). A loophole that projects out from an opening in the timber of a keep or tower, that provides a clear shot for a defender downwards and left or right at attackers at the base of the wall.

See *'sama', 'tenshu', 'yagura'*.

Diao qiao. (Ch. *angling bridge*). A drawbridge, built at the narrowest point of a moat, or two drawbridges are used to span the moat connected by a fortified artificial islet at the midpoint of the moat. *'Diao qioa'* can also refer to a suspension bridge.

Diatenshu. (J. *great castle tower*). A large *'tenshu'*, with at least five tiers and six stories.

See *'tenshu'*.

Dike, dyke. (O.N. *dik*, dam). A long linear work consisting of a ditch and rampart denoting a boundary. Examples are Offa's dyke (Wales), and Denverk (Germany).

See 'ditch', 'dyke', 'moat', 'rampart'.

Di lou. (Ch. *opposing tower*). Normally a permanent tower built on the wall top or bastion, with a sloped roof and awning, to protect defending troops, could be temporary or semi-permanent which would be called a *'zhan peng'*. Also known as a *'jian Lou'* (*'arrow tower'*).

Diminished angle. The angle formed by the front and the exterior side of a bastion.

Direct fire. To fire directly at an enemy position or fortification.

Directorate of Fortifications and Works. ***See 'FW3'.***

Directrix. (Fr.).

(1) The first line traced on the ground in laying out a fortification.

(2) The centre line of the plane of fire for cannon in an embrasure or platform.

See '*embrasure*'.

Dirty tricks, Playing dirty. Warfare is not played nicely and dirty tricks were used, including threats to a castle's garrison were real, including:

- William the Conqueror ordered that the hands and feet be cut off the remaining garrison at Aleçon (France, 1047 CE);
- Theobald I in Italy had the captured garrison castrated (10th century CE);
- Crusading Franks set fire to sacks and cushions and threw them over the walls of fortifications to act as a '*smoke screen*';
- Offering bribes of positions or monies to traitors to let the opposition into the fortification or provide crucial information;
- Sneaking into a fortification disguised as foragers to gain access or just to spy;
- During the siege of Constantinople by the Turks in 1453, prisoners were impaled on wooden stakes.

Disappearing turret. ***See 'eclipsing turret'.***

Dismantlement. The deliberate deconstruction of a fortification that is no longer required or meets the current needs.

Di tai. (Ch. *opposing platform*). A tall fortified structure with a flat roof and battlement, be it a wall tower, bastion, tower not connected to a wall. Also known in the Song period as a '*nu tai*' (Crossbow platform). Di tai came in two types based on construction:

(1) '*Shi xin di tai*' (Ch. *solid opposing platform*) is a solid structure.

(2) '*Kong xin di tai*' (Ch. *hollow opposing platform*) is a structure provided with rooms for troops to fight from. Normally towers taller than adjacent walls would be hollow and the bastion at the same height as the adjacent walls would be solid, but this could vary.

Ditch. (O.E. *dic*, dike). A wide trench excavated along the outer perimeter or between parts of a fortification, which was utilised to impede the approach of an enemy force towards the walls thus reduce the effectiveness of siege weaponry, such as siege towers and battering rams, which required close proximity to the walls to be effective. The ditch was either filled with water or left dry. And if the ramparts were earthen then the soil removed from the ditch was used to raise the ramparts.

See '*dike*', '*cunette*', '*double ditch*', '*front ditch*', '*moat*', '*rear ditch*'.

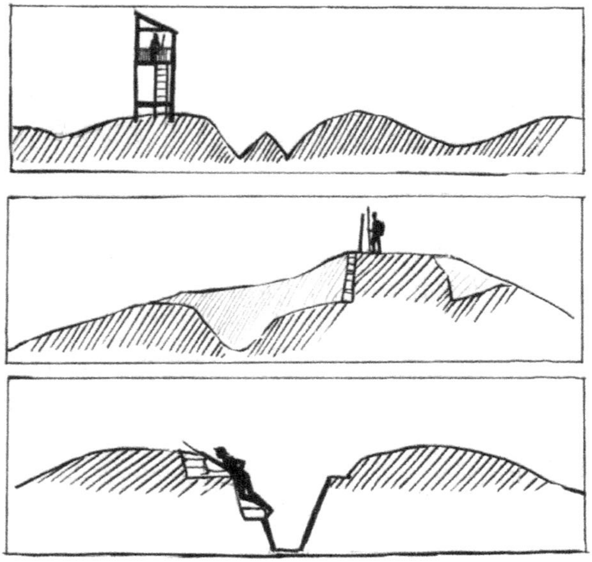

Ditch profiles; Roman, Iron Age hillfort, British WWI trench.

Di tuan. (Ch. *opposing round*). A round or rounded bastion. Also known as '*tuan luan*' (round tower).

Dobashi. (J. *earthen bridge*). An earthen bridge left intact when excavating a moat in front of a castle entrance, which acts as a dam for traffic of boats in the moat or in the case of a dry moat ('*karabori*'), the movement of enemy troops in the moat.

Dodecagon. A fortification of twelve sides.

Dobei. (J.). A wall made by daubing clay over a reed or bamboo framework and topped by a small tiled roof. The clay was often mixed with grass ('*wara*') to add reinforcement. In the Edo period the dobei was covered in white plaster to increase strength and improve weather resistance.

This type of wall was used in two different parts of a Japanese castle:

(1)the dobei were built along the top of the '*ishigaki*' (or the main walls) linking the corner and connecting towers, providing protection to the defenders within;

(2) the dobei were also raised on both sides of the castle's major roads, effectively screening the rest of the compounds from enemy eyes, and they formed a maze which hindered the enemy greatly if they gained entry.

Other variations of the dobei were constructed in different ways and were known as *'neribei'* and *'tuijibei'*.

See *'hikaehashira', 'karawa dobei' 'neribei'* and *'tuijibei'*.

Doi. (J.). The earthen rampart on the outside of a moat surrounding a castle, made from the spoil when digging the moat. Other defensive installations associated with the *'doi'* included; palisades (*'saku'*), surrounding enclosures (*'kuruwa'*), stone faced walls (*'ishigaki'*), gates and entrances (*'koguchi'*). Dirt excavated to form the *'doi'* could be made by un-loading the spoil in one place or placing the dirt in wooden frames called *'hanchiku'*. applying compaction pressure and water to make the rampart stronger, called *'tatakidoi' with a gentle slope. Whereas, a 'doi' with a grass embankment and a steeper slope is called* 'shibadoi'. The parts of a *'doi'* are the base (*'doijiki') and* the flat top (*'mabumi').*

See *'choujou', 'dorui'*.

Dokuritzushiki. (J.). A *'tenshu'* that stands independent of other *'yagura'*.

See *'borogata', 'boroshiki tenshu', 'fukugôshiki tenshu', 'renketsushiki tenshu', 'reritsushiki tenshu', teiritsushiki tenshu'*.

Dom bunker. A concrete bunker with walls 80 metres long and 2.5 metres thick (262 X 8.2"), used by Germany in WWII to house artillery on rails or submarines. Also known as a *'cathedral bunker'* due to its pointed arch shape. A number of such bunkers can be found in Calais and Lorient, France.

See 'bunker', 'U-boat pen'.

Dombunker for the German railway gun Krupp K5 (E) near Hydrequent, France.

Dome. Any type of armoured turret.

See *'cloche'*, *'turret'*.

Donjon. The main defensible building of a fortification (the main tower or citadel), which was used as the Lord's residence or as the last refuge if the outer defences fell to the attackers. The word is derived from the Latin word for lord *'dominus'*. The term was later corrupted to *'dungeon'*, as prisoners were often kept in the lower levels of the tower. Also known as the *'keep'* or *'great tower'*.

See *'keep'*, *'mote'*.

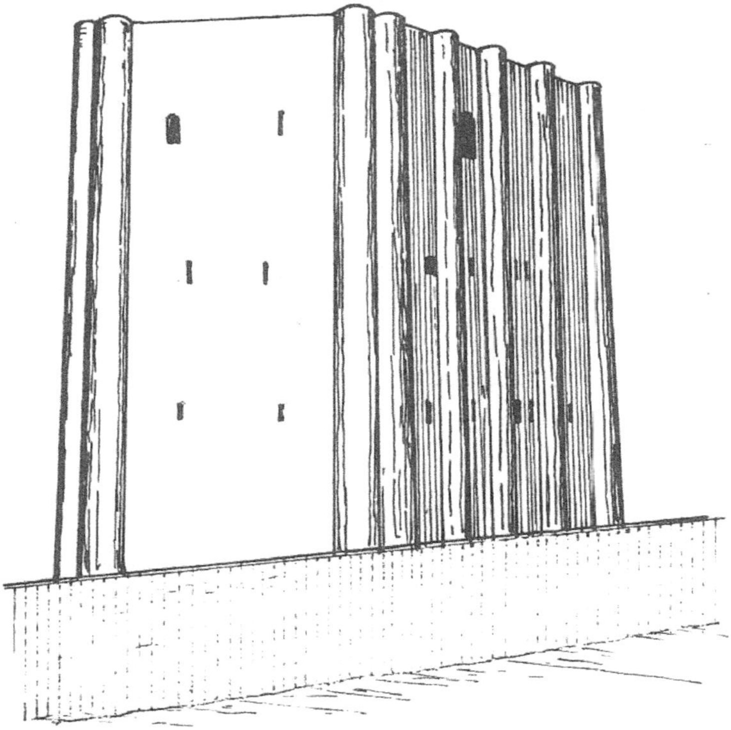

The donjon at Langeais (France) (above) was raised by Fulk Nerra (the Black) in 995 CE consisting of a rectangular tower 16m x 7m (52 feet x 22 feet) in plan, entry was made by a small turret, stairs from ground to 1st floor living area. The ground level was used as a storeroom. This is ascribed as the first castle of the medieval period.

See 'tower', 'keep'.

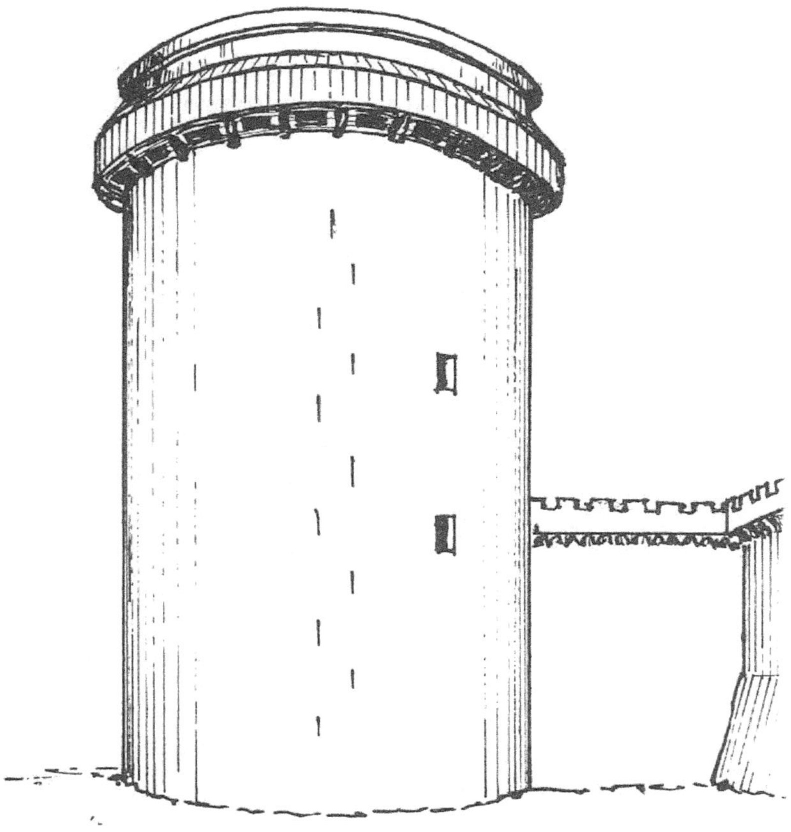

The donjon of the Château de Falaise is a castle from the 12th-13th century, located in the south of the commune of Falaise in the département of Calvados, in the region of Normandy, France.

Door beam. A beam for securing the doors of a fortification. The beam is usually fitted into sockets on either side of the entrance and into brackets on the door itself.

See 'put log'.

Dorui. (J.). The earthen rampart of a castle.

See *'doi'*.

Dôshinén. (J.). One of the three major types of Japanese castles which date from around 1600 CE, constructed using a plan of three fortified compounds arranged in concentric pattern; the *'honmaru'* in the centre was surrounded by the *'ninomaru' which* in turn was surrounded by the *'sannomaru'*. Variations on this occurred depending on the site.

Dote. (J.). The flat top of an earthen embankment (*'doi'*). Also known as a *'mafumi'*.

See *'doi'*, *'nagadote'*, *'kazashi no dote'*, *'mafumi'*.

Double castrum. Concentric fortification, introduced in the 13th century CE in Europe.

Double ditch. A parapet with an interior and exterior ditch, used to widen an existing *'back ditch'* parapet or to provide a *'covered way'* behind the parapet of the *'front ditch'*. If an earthwork was captured it could be turned and would have a *'double ditch'*.

See *'ditch'*.

Double hornwork. An advanced work consisting of two hornworks, one in front of the other.

See 'hornwork'.

Double tenailles. Part of the Dutch fortification system using tenailles.

See *'tenaille'*.

Double cross gun loop. Similar to the *'cross and orb gun loop'* but the vertical slit rose approximately twice the height and was provided with cross slits, one at the top, and the second a third of the way down.

See 'cross and orb gun loop', 'gun loop'.

Dou yan. (Ch. merlon hole). ***See 'tian jing', 'xuan yan', 'yan'.***

Dou yan (in the battlements) at the Nanming Gate, Lishui City, China.

Dover quad pillbox. A British WWII pillbox built in the Dover area to protect the port of Dover (England), often found in commanding positions, square in plan with wide embrasures on all sides for rifle or machine gun defence, with an overhanging roof, there are twenty four examples remaining, an example can be found on the Western heights, Dover, England.

See 'British hardened field defences', 'FW3', 'pillbox'.

Dôzo-zukuri. (J.). Since most of the buildings of a Japanese castle were mainly constructed of wood there was the necessity of reducing the risk of fire. This was accomplished by the technique of covering the walls and timbers with thick layers of plaster, which is known as '*dôzo-zukuri*'. The plastering was used as a protection against domestic fires as well as those caused by the incendiary missiles of an enemy.

Drahthinbernis. (G. Barbed wire obstacle).

See '*barbed wire*'.

Drainage ditch. See '*cunette*'.

Dragon's teeth. An Anti-tank obstacle, set out in belts of four to five rows of reinforced concrete pyramid shaped blocks, varied in height from 80 cm (2 feet 7 inches) to 1.4m (14 and ½ feet). Used by the French (Maginot line), German (Siegfried line and Atlantic wall), Russia (Ukraine/ Russia War).

See '*anti-tank pimples*', '*denti di drago*'.

Draw bolted doors. Doors which were used to defend an entrance of a fortification, secured by draw bolts, such as those of a forebuilding of a keep.

Drawbridge. (L. *pontem torneicuim*). A bridge which was used to provide access to a fortification, and when in the raised position it closed the entrance. Generally, a drawbridge was hinged at the bottom and free at the top, and could be drawn up to prevent an enemy gaining entry. The drawbridge usually spanned a ditch or moat, or the part of a

ditch or moat between the fortification and a causeway. Its simplest form consisted of a movable plank; others were pulled up by chains worked by pulleys or a windlass.
A later development was the *'hinged platform'*; which could be raised by pulling up chains attached to the outer corners, these chains passed through slots above the entrance and were attached to a windlass in the chamber above the entrance. The most elaborate type worked on a counterpoise system; the chains were suspended from beams which, when the bridge was drawn up, fitted into recesses provided above the entrance. Another type worked on the pivot principle, where the inner part of the bridge was moved into a pit while the outer part rose to completely cover the entry, also known as a turning bridge.

See 'bascule bridge', 'bridge pit', 'hinged bridge'.

Drop box. A feature in gatehouses and barbicans, consisting of holes through the floor of the structure over the entry, that were used to drop missiles and other offensive materials on the enemy below.

See 'meurtriéres', 'murder holes'.

Drum tower. A round tower which was used as a flanking tower for curtain walls, gatehouses and barbicans.

See 'flanking tower'.

'D' shaped flankers. A flanking tower which was used to defend the corners of a keep, consisting of a cylindrical tower with the interior side cut square, producing the *'D'* shape.

See 'flanking tower'.

Duck boards. Trenches dug into the ground often fill with water. To prevent soldiers feet from getting wet and developing *'trench foot'* trenches were provided with a sump and above that level the *'duck boards'*.

See *'trench'*.

Duelling field. See *'pré carré'*.

Dug out. A hole dug in a side of a trench, usually sited in the rear of the support trench, shored up with timber and or corrugated iron. Used to provide soldiers with some protection during their time in the trench, including when waiting, eating and sleeping. Also known as a *'bolt hole'*.

See *'trench'*.

Dumble tank stops. (E. *dumble*, stupid). Vertical timber posts were set every 1.5 m (5") along the northern bank of the ditch of the Australian WWII (1942 CE) Belmont (New South Wales) Anti-Tank Ditch.

See 'Belmont anti-tank ditch'.

Dummy trenches. Trenches which simulate real trenches, but are used to deceive the enemy as to the location and number of defending troops, located to the rear. Used in the Boer war, and by the Japanese with their war with Russia.

See *'trench'*.

Dummy gun pit. A deception consisting of a mound of earth with a log protruding out over the crest, which would draw enemy artillery fire.

See *'gun pit'*.

Dump or Glacis rampart. The simplest method of constructing a rampart, by excavating a ditch and casting the soil inwards to form a *'dump'*. Often reinforced by a wall or palisade along the top.

See *'glacis'*, *'rampart'*.

Dun, Dún, Dúin, Dhúin. (Gaelic *dún*, fortress). A fort dating from the late Iron Age, which is found in the west and the south-west of Scotland. It provided protection for a family or a small group. Similar in construction to a broch, a dun had a diameter of about 18 metres (59 ft), and a narrow doorway angled through the dry-stone wall, the walls were up to 4.5 metres (15 ft) thick. Usually circular or *'D'* shaped, they were frequently built on coastal headlands. The dun was different from hill forts mainly because of their smaller size and the complexity of their internal lodgings. Note: It appears in some text that *'dun'* and *'broch'* are substituted for each other.

See broch, hill fort.

Duo kou. (Ch.). A crenel of a parapet. Also less commonly called a *'nu kou'*.

Durer, Albrecht (the Younger). (1471 – 1528 CE). An artist born in Nuremberg. Durer also was a writer best known for *'Instruction of the fortification of towns, castles and large villages'*, published in 1527 CE.

Durg, Durga. (Sanskrit, *fort*). A Indian fort which can be differentiated by how it was defenced, its location, be it on a hill top, in a valley, on a plain, in a marsh or surrounded by water, in a desert, in a forest, or by what it was made from (mud, earth, brick or stone).

See 'airina-durga', 'antardvipa-durga', 'giri-durga', 'guha-durga', 'jala-durga', 'khanjana-durga', 'mahi-durga', 'mrid-durga', 'nara-durg', 'nri-durga', 'panka-durga', 'parigha-durga', 'stambha-durga, 'sthala durga', 'vana-durga.

Dutch Water Line. (Du. *Holland Waterline*). A Dutch system designed in the early 17th century CE, combining inundation and fortifications, consisting of sluices in dikes, forts, batteries and fortified posts. The Dutch Water Line was updated in the 19th and 20th centuries CE (85km, 53 miles in length) but was no defence against the German *'Blitzkrieg'* in 1940 CE.

See 'Grebbe line', 'Vesting Holland'.

Dwingel. A bent barbican dating from the 14th century CE.

See 'barbican'.

Dyke. (O.N. *dik*, dam). A linear defensive earthwork which covered long distances, consisting of ditches and ramparts. An example is Offa's dyke which roughly follows the border of England and Wales.

See 'dike', 'ditch', 'moat'.

Dzong. (*Dzongkha fortress*). A fortified type of monastery, found mostly in Bhutan and Tibet, with tall exterior walls surrounding a complex of courtyards, temples, offices and accommodation.

E.

Ear. ***See 'bastion', 'flank', 'orillon'.***

Eared pillbox. A British WWII pillbox design in 1940 CE, located only in the north east of England, irregular hexagon in plan, with two large forward facing embrasures at 180^0 to each other, for medium machine guns, but no loopholes for rifles for close defence, normally located near beaches, a short anti-ricochet wall extended perpendicular from the rear wall, two entrances were provided on either flank facing forward. There are 16 such pillboxes extant. An example can be found in Speeton, North Yorkshire, England.

See 'British hardened field defences', 'FW3', 'pillbox'.

Early Warning Stations. Stations using radar and other devices to detect and provide warning of attack by aircraft or intercontinental nuclear missiles, so people could evacuate to the safety of a shelter.

Eastern command pillbox. A British WWII pillbox variant is an infantry version of the '*Vickers Medium Machine Gun emplacement*' with embrasures for Bren guns in the front faces and to one side of the entrance, the entrance was protected by thick blast wall. Common in Essex and Suffolk, there are sixty four extant examples, such as the pillbox near Rettendon in Essex, England.

See 'British hardened field defences', 'FW3', 'pillbox'.

Earthwork. A fortification which was made chiefly of earth; either for temporary or permanent use, for either defensive or offensive purposes, constructed by excavating and embanking earth.

See* 'glacis', *'rampart'.

Earthwork batteries. Batteries made of earth. The earth of the earthworks absorb the impact of cannon fire, where masonry would shatter. Earthworks were quick and cheap to throw up but required constant maintenance.

The English raised earthwork batteries in coastal towns during the Hundred Years War with France in the 14th century CE. At Calais in 1436 CE the English erected bulwarks of earth to defend against attacks by the French.

Earthwork complex. A complicated earthworks consisting of fortified fronts between two armies, consisting of the '*zone of contention*' or '*no man's land*', then a '*zone of fire*', with advanced positions of the opposing combatants, fronting the 'main line of defence', followed by a '*zone of occupation*'.

See 'earthwork', 'no man's land', 'trench warfare'.

Echaugnette.
(1) A watch tower or bartizan.

See 'bartizan', 'machicolated turret'.

(2) A round sentry box sited at the angle of a bastion, which was corbelled out so as to facilitate observation of the surrounding terrain.

Eclipsing turret. An armoured retractable turret, armed with one or two light guns, developed in the 19th century CE and deployed on the Maginot Line. The eclipsing turret was expensive and only lightly armed, and the mechanism for lifting and lowering was un-reliable.

See 'turret'.

Ecoute. (Fr. *listen*). A small gallery situated in the foreground of the glacis, connected to a main gallery running parallel to the covert way, where a miner would listen for the noises made by enemy miners, thus enabling countermines to be dug to intercept them.

See 'catacoustics'.

Eckturm. (G.). Corner tower.

Edwardian castle. The castles which were raised by Edward the 1st of England, to control the newly acquired territories in Wales, these castles are amongst the finest examples of concentric fortifications of the period.

Photos of various Edwardian castles are provided in the centre of the book, including: Beaumaris castle (gateways, moat, ward); Conwy castle (enceinte); Harlech castle (gatehouse).

See 'concentric castle', 'great gatehouse', 'ring of iron'.

Eglise donjon. (Fr. *church tower*). A fortified church or chapel.

Egyptian merlons. Half circle merlons used on Egyptian fortifications (i.e. *Medinet Habu*, the palace of Rameses III, about 1200 BCE).

See 'merlon'.

Eingangsverteidigung. (G. *entrance defence*). A small firing hole closed by a sliding metal shutter, used to defend the entrance to a German bunker, used in both WWI and II.

See 'bunker'.

Eisenbeton. (G.). German reinforced concrete with iron or steel, steel replaced iron in about 1920 CE.

See 'stahlbeton'.

Elastic defence. *See 'defence in depth'.*

Elbow rest. Part of a trench where a soldier rests their elbows when standing on the fire step, while holding a rifle to shoot above the parapet.

Electric Searchlights. Used to highlight enemy aircraft or shipping. Example is the Needles Battery, Isle of Wight (1889-1892 CE) in association with the Quick Firing Gun batteries.

Element cointet (Element C). ***See 'beach obstacle', 'Belgian gate'.***

Embattled. ***See 'battlemented', 'crenelated'.***

Embrasure, Embrazure. (Fr. *braser*, to splay).

(1) An opening in the parapet or the wall of a work for firing guns through at an enemy. The embrasure was internally splayed to allow the gun to be swung through a greater arc, thus increasing its field of fire.

(2) The splayed interior of a gun loop, gun port, arrow loop or slit.

(3) A crenel or a reveal.

See *'cheek', 'floor', 'loop hole', 'throat'.*

Embrasure battery. A battery that fired through an embrasure in a parapet wall, and thus provided some protection to the cannon and its crew.

See' *barbette', 'battery'.*

Embuscade. (Fr. *ambush*). A concealed place in a fortification used to attack an enemy by surprise.

Emergency coast batteries. Following the fall of France in (1940 CE) England set up artillery batteries in quick setting cement provided with long range guns. Coastal defence had been ignored until then. The batteries of Dover and other major ports, later extended from Lincolnshire to Cornwall.

Empty bastion. When the level area within a bastion is lower than the ramparts the bastion is said to be empty.

See *'bastion'.*

En barbette. When cannon are mounted on high platforms or carriages so they can be fired over the parapet rather than through an embrasure.

En bec. A tower which is described as being '*en bec*' is a tower which has a beaked projection pointing towards the most likely direction of attack by an enemy's attack. Examples of '*en bec*' towers can be found in France at Carcassonne (Narbonne gate), Loches (curtain wall) and the Gate of St. John at Provins.

See 'ogival point'.

Enceinte. (Fr. *enceindre*, to surround).
(1) The body of a fortification, a fortified enclosure, generally the whole area which was enclosed by the main wall or rampart.
(2) A fortified perimeter.

En cremaillere. A work with a series of offsets or notches.
See *'cremaillere'*.

Enfilade. (L. *flium*, thread). Defensive artillery was fired from the flank of a work and directed along or across another, for example; from the salient of a bastion across the faces of an adjoining bastion or the curtain wall in between. Enfilade fire produces a better form of defence because it is more effective against advancing infantry than direct fire.
See *'fire'*.

Engineers (Siege). Builders, maintainers and users of siege engines were also known as '*machinators*' or '*ingeniators*'.

Enneagon. A nine-sided fortification.

Ensconce. To fortify, entrench, or protect with fortification.
See *'inskonse'*.

Enshõgura. (J.). A gunpowder magazine, made with thick stone walls.

Entanglements. Barbed wire entanglements, used to control access at frontiers and defend ditches, trenches and other fortifications.

See 'barbed wire', 'high wire entanglements','low wire entanglements', 'spider wire entanglements'.

Entrée. (Fr. *entrance*). The different types of entrances to the Maginot Line were;
(1) Entrée Hommes (EH). The entrance for the troops.
(2) Entrée Mixte or Entrée et Hommes et Munitions (EMH). The entrance for troops and munitions.
(3) Entrée Munitions (EM). The largest of the entrances to take trucks and the narrow gauge railway for the entrance of the munitions.

Entrench. To fortify with a trench and parapet.

See *'parapet', 'trench'.*

Entrenched bastion. A type of bastion, which could be used as a '*reduit*' or secondary citadel because it was both self-contained and defensible fort, with an enclosed wall, a moat and drawbridge in the gorge.

See 'bastion'.

Entrenched camp. ***See 'camp retranché'.***

Entrenching tools. E-tool. Tools used for digging field fortifications, be it a small collapsible spade carried by soldiers or spades, pickaxes and axes carried on the outside of tanks or by Roman soldiers.

See *'sappers'.*

Entrenchment. (It. *trincieramento*). A defensive earthwork consisting of trenches and parapets.

See *'parapet', 'trench'.*

Entrenched camps. Protected areas for the safe assembly or regrouping of an army behind a permanent fortification, this may be in a town or a city.

Envelope. (Fr.).
(1) A work which covers or defends another work.

See *'demilune', 'épaule', 'pâté', 'tenaillon'.*

(2) A small rampart built in a ditch or in front of it, which was used to cover a weak point in the works.

Épaule. (Fr. *épaule*, shoulder). An earthwork situated where the flank and the face of a bastion join to form the shoulder of the bastion, used as a cover from flanking fire.

Epaulement. A side work of a battery or earthwork used to protect from flanking fire.

Episcopal castle. (Gk. *episkopos*, overseer). A castle which was governed by a bishop.

Eptagon. A fortification which has seven sides.

Escalade. A form of attack that requires the climbing of a fortification's walls by the use of ladders.

See Appendix 3. Siege Warfare - The art of Offence and Defence.

Escaraguaita. A Spanish hexagonal shaped bartizan of the 14th century CE.

See 'bartisan'.

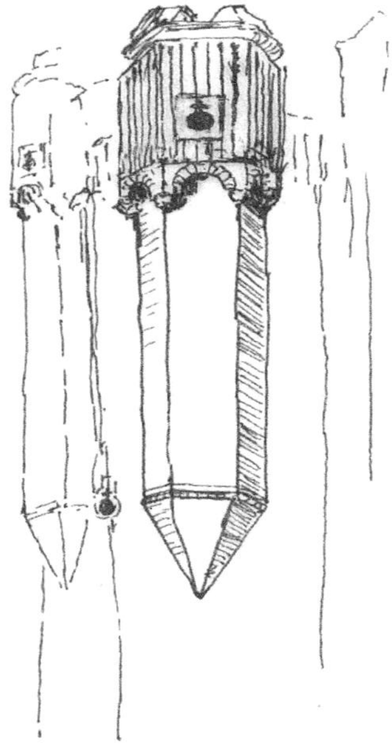

Escaraguaita with battlements, orb and cross gun port, corbelled out from the tower. 15th century CE Castle of Coca is a castle located in the Coca municipality, central Spain.

Escarp.

(1) The side of a ditch which was next to the rampart.

(2) To make into a scarp or sudden slope.

See 'scarp'.

Escarpment. The foreground of a fortification, which was excavated precipitously to hinder an enemy's approach.

Esplanade. The space between a citadel and the surrounding houses of a fortified city, which was purposely kept clear so as to eliminate any cover for an enemy force attempting to erect breaching batteries or making approaches.

Essex lozenge pillbox. A British WWII pillbox built into the sea walls (erosion protection) of Essex, are elongated octagon in plan, with three embrasures on each end and two in each side, the door is on one of the sides has a '*L*' shaped porch, a '*T*' shaped anti-ricochet wall runs the length of the pill box with the top of the '*T*' facing front. There are 36 such pill boxes extant.

See 'British hardened field defences', 'FW3', 'pillbox'.

Estacade. (Sp. *estaca*, stake).

(1) A barrier or dike constructed of wooden piles driven into the sea or river bed, to obstruct an enemy's advance.

(2) A form of defence made of stakes, such as a stockade.

(3) A raft fastened together with chains, which was used to render a channel or harbour mouth impassable.

Etoile. (Fr. *étoile*, star). A small star shaped fort or redoubt which had four, five, six or more points, also known as a '*star redoubt*'.

Exagon. A fortification which has six sides.

Expanded metal plate (XPM). Metal sheet with lines of slits cut into the sheet, then stretched, opening up the slits. Used to strengthen ('*revet*') trench walls. Other material used was; corrugated tin sheet, sand bags, hurdles made from woven branches.

See *'revetting'.*

Expense magazine. See *'powder magazine'.*

Exterior crest. The line of intersection of the superior and exterior slopes.

Exterior ditch. The ditch on the outer side of the parapet, between the attacking enemy and the parapet.

See*'ditch', 'front ditch'.*

Exterior side. The side connecting the exterior crest and the berm.

Exterior side of a fortification. The exterior side of a fortification is the distance or the line between one bastion and the next.

External polygon. The line connects the point of an arrow headed bastion to the adjacent rampart, making a 45-degree angle with the centre of the bastion.

See *'arrow headed bastion', 'internal polygon'.*

Exterior slope. The outer side of the parapet, from the '*superior slope*' to the '*berm*'.

See *'berm'.*

F.

Face.

(1) The front between two neighbouring bastions or any other type of salient work.

(2) The part of a work formed by one side of a salient angle.

(3) The outer surface of any work projecting forwards which meet to form a salient angle.

Facing. The covering of a slope of a work with sod, sandbags, stone or other material to strengthen the slope.

See *'revetment'*.

Face of a bastion. The exterior face of a bastion between the salient point and the flank.

See *'face'*.

Fahrpanzer. ***See 'panzerlafette'.***

Fall trap. A form of gateway defence. **See *'drop box'*, *'meurtiéres'*.**

Fallgitter. (G.). **See *'porticullis'*.**

False machicolations. Brackets serving a purely decorative function.

See *'blind machicolations'*, *'bracket'*, *'mock machicolations'*.

Fakir's bed. A plank of timber with exposed nails, used to control traffic access as well as being an anti-personal obstacle (Berlin Wall, East Germany).

See 'Anti-personal obstacles'.

Fangrost. (G. *catch grate*). Re-enforced concrete beams set above German *'U-boat pens'* to absorb the explosive force of bombs dropped by allied bombers during WWII.

See 'U-boat pens'.

Fascines. (L. *fascis*, bundle). Bundles of brushwood were used to; fill in ditches, used as *'revetment'* in the construction of field works, strengthening earthworks and to cover wet ground and prevent erosion. There were three types of fascines; 1) *'battery or long fascines'* (*'saucissons'*), 2) *'trench fascines'* and 3) *'water fascines'*.

Fascine choker. A device used to tighten fascine bundles consisting of two long wooden poles connected by chain, the chain was looped around the saplings and using the leverage of the poles to tighten the bundle. The fascines were then secured by withes or gads (willow sapling twisted to tie in knots).

See *'fascines'*.

Fascine horse. A device used to hold saplings in place to make into fascines.

See *'fascines'*.

Fasil. (Ar.). A low third wall outside the double walled defences of a town. Similar to Constantinople's *'Theodosian land walls'*.

Fastness. (O.E. *fads*). A major fortification; a stronghold or fortress.

Fausse braie, Fausse braye, Fausse bray. A low rampart provided with a parapet built forward, parallel and below the main rampart, forming an outer enceinte, used in the 16th and 17th century CE by engineers as a covered way. Later the work was abandoned because when under bombardment the fragments from the rear wall wounded the defenders present, forming what came to be known as a *'shot trap'*. M. Vauban continued to use these types of works placing them before curtains, they were then called *'tenailles'*.

See **'*tenailles*', '*Vauban*'**.

Fence (fortified). A fence used to designate or defend an area or border. Examples are the Hungarian/Serbian or USA/Mexico borders.

See **'*Berlin Wall*'**, 'dacoit fence'.

Fer a cheval. (Fr. *horseshoe*). A work with a parapet in the shape of a horseshoe.

Fernkampwerk. (G.). An artillery position behind an infantry position, part of an Austrian strong point system, in 1900 CE.

Ferte. Fort (German from Latin *firmitas*, firmness).

Feste (G.).

(1) Decentralized groupings of mutually supporting forts and batteries, with barracks, trenches, barbed wire entanglements (1800's CE).

(2) A stronghold. A large and heavily defended fort to the 1890's, later Feste had dispersed position in irregular shapes.

Festung. (G. *fortress*). A fortress, which could include multiple forts and/or a *'feste'*.

See **'*feste*'**.

Festung Norwegen. (G. *Fortress Norway*). The defensive fortification system constructed by Germany during WWII in Norway as part of the Atlantic Wall.

See **'*Atlantic Wall*'**.

Field engineering. The planning and practice of making temporary military fortifications and support structures like roads and bridges during an operation.

See **'*field work*'**.

Field gate. A gate in the curtain wall leading out to the country surrounding a castle.

See **'*postern*'**.

Field gun emplacements. A British WWII emplacement for protection of field guns, similar to Pill box 28, about 50 are still extant.

See 'British hardened field defences', 'FW3', 'pillbox'.

Field magazine. A structure used to store gun powder on the flanks of an artillery position or dug into the parapet.

See *'powder magazine'*.

Field fortification, field work. A temporary work constructed by an army in the field, used to cover an attack on a fortification, strengthen a current fortification, or as protection against another enemy army, especially a relieving force.

Fighting hole, fighting position, fighting pit. See *'fox hole', 'tobruk'*.

Filling room. A small powder magazine where prepared ammunition (i.e. a bullet rolled up with gunpowder to make a cartridge) is stored close to where soldiers were stationed.

See *'powder magazine'*.

Fire. Weapons fire can be direct when it comes from the front, oblique or slanting with it strikes at a small angle, enfilade if it comes from the side, or reverse if it comes from the rear of a defended work.

Fire bay. A straight section of a trench with a parapet from which defenders can fire forward at the enemy, with protection on the side was protected by earthen rampart and sandbags.

Fire control tower. A tower located near a coastline, used to detect and locate enemy shipping, to direct fire from coastal batteries, and adjust the aim by providing corrections. Used by a number of countries for coastal defence.

See *'leitturm'*.

*German **fire control tower** of type 'Marine Level 3' was part of the 'Atlantic Wall' on the island of Alderney.*

Firing crest. The upper part of a breastwork over which soldiers could fire.

See 'breastwork'.

Firenzuolo lines. *See 'Santa Margherita Lines'.*

Fire support base. Artillery strong point used by USA troops during the Vietnam War dominating the surrounding area, consisting of sandbagged and dug in emplacements for guard posts, machine guns, recoilless rifles, grenade launches, mortars, howitzers, infantry bunkers, plus command posts, communication post, fire control and observation tower. Surrounded by a barbed wire perimeter fence, claymore mines.

Fire step. Used by soldiers to step up to the parapet and fire a weapon over a parapet be it bow, crossbow, arquebus, musket or rifle. Examples are: Bedzin Castle, Poland 13th CE. A fausse braye parapet, late 16th century CE.

Fire trench. A trench deep enough for a soldier to stand in.

Firing gallery. A gallery which was set into the walls or towers of a castle below the level of the battlements. A firing galley was provided with slits for firing arrows and bolts at an enemy. With the introduction of the galleries the fire power of a castle's defences was increased.

See *'mural gallery'*.

Firing loop, Firing slit. Apertures in the walls of a castle which were used by defenders to discharge arrows and bolts though at the enemy.

See *'arrow loop'*, *'arrow slit'*, *'loop'*.

First fortified belt. Part of the French Maginot line, consisting of a line of fortifications in front of the Maginot Line supported by small redoubts and pill boxes armed with anti-tank weapons and machine guns, along with some cupolas with short barrelled mortars.

Flak tower, Flakturm. German WWII tall towers with thick walls and mounted by flak cannons as air defence, with enough surplice space for public air raid shelters and hospitals. Flak towers were also set up in the Thames and Mersey Estuaries by the UK.

Flakturm. *See 'flak tower'*.

Flame thrower.

(1) A siphon used to project *'Greek fire'* by the Byzantines to attack enemy ships.

(2). A backpack or tank mounted flamethrower used to clear tunnels and other fortifications of an enemy, used by the USA in WWII.

Flanderzaum. (G. *flandrian fence*). Double apron barbed wire fence.

See 'barbed wire'.

Flank. (Fr. *flanc*).

(1) The flank or side of a work, such as that of a bastion from the salient point to the curtain wall, used to provide a position for flanking fire.

(2) To provide defensive flanking fire which defends another work along the length of the work.

Flank (concave). The flank which is made in an arc of a circle.

Flank (direct or grazing). The flank which is perpendicular to the opposite face produced, and oblique when the flank makes an acute angle with that face.

Flank (retired). The flank which is built behind the line of the face of the bastion and the curtain.

See *'retired flank'*.

Flank (second). The flank which is made when the face of the bastion does not extend to the curtain but joins it at some other point, then the part of the curtain between that point and the flank is called the second flank.

Flanked angle. The angle formed by a curtain wall and a flank of a bastion.

Flanker.

(1) A work which commanded the flank of an assailing force.

See 'flanking', 'flanking tower'.

(2) A battery in the flank of a bastion, which was used to flank the adjacent rampart.

Flanking. The use of fortifications designed to provide flanking fire, for example; the approaches to a tower were covered by the neighbouring curtain walls. Flanking reduced the amount of dead ground surrounding a fortification thus improving its defensibility.

See 'dead ground', 'flank', 'flanking tower', 'yokoyagakari'.

Flanking angle. The angle formed by a flank of a bastion or other work with a curtain wall. Also known as a *'curtain flank'*.

See 'curtain angle'.

Flanking effect. The effect of flanking fire on an enemy or their position.

See 'maximum flanking effect'.

Flanking tower. A tower which was built not only beside the entrance but also at strategic points on the curtain wall, usually at the angles, which were used to provide flanking fire.

Flanking turret. A turret situated at the corner of a castle which was used to provide flanking fire.

See 'bartizan', 'machicoulis'.

Flanking wall. A curtain wall flanking a tower or bastion or some other work.

Flank of the bastion. The section of the bastion between the face and the curtain, from which the ditch in front of the adjacent curtain and the face of the neighbouring bastions were defended.

See 'enfilade', 'flank', 'flanking'.

Flat bastion. A bastion where its two faces form an angle with little or no salient (i.e. Flat bastion at Gibraltar).

See 'bastion'.

Fléche. (Fr. *fléche*, arrow). An arrow shaped outwork consisting of two faces with a parapet and an open gorge, which formed a salient angle at the base of a glacis, smaller in size than a *'redan'* or a *'lunette'*.

Fleche-redan. A small redan built at the base of a glacis of a bastion or a re-entering place of arms.

See *'bastion', 'glacis', 'place of arms', 'redan'.*

Fliehburg. (G. *refuge castle*). A defensive location surrounded by ramparts, not permanently occupied but acts as a temporary refuge in times of war or attack. Also known as *'fluchtburg','volksburg'*, *'bauernburg'* or *'vryburg'*.

See *'castle'.*

Floor. The bottom part of a loop hole or embrasure, also known as the *'sole'*.

See *'cheek', 'embrasure', 'loop hole', 'throat'.*

Floriana Lines. (Maltese. Is-Swar tal-Furjana). A line of fortification consisting of trace italiane with a land front with four bastions, two cavaliers, curtain walls and a deep ditch surrounding Valletta, and forms the outer defences. Construction began in 1636 CE, modification in the 17th and 18th centuries CE, and saw use in 1798 to 1800 CE during the French blockade.

See *'Corradino lines,' 'Cottonera lines', 'Santa Margherita Lines'.*

Flying bridge.

(1) A narrow wooden bridge which was used for the purpose of communication between the motte and the bailey of a *'motte and bailey castle'*. The flying bridge was usually supported on wooden piles, the

bridge could be then demolished by the defenders if the enemy took the bailey.

See *'motte'*.

(2) A small bridge which was used for the purpose of intercommunication between the different parts of a fortification (e.g. between outer and inner defences). They could be withdrawn, thrown down or destroyed if necessary, depending on the situation. Example is the motte and bailey castle depicted in the Bayeux tapestry

See *'chemise foot bridge'*.

Flying sap. The rapid construction of a siege trench.

See *'sap', 'siege', 'trench'*.

Foot of glacis. The line where the slope of the *'glacis'* meets the level ground of the open land in front of a fortification.

***See 'glacis'*.**

Forebuilding.

(1). A subsidiary tower projecting in front of a castle's main entrance.

***See 'barbican', 'gatehouse', 'small keep'*.**

(2). A small tower which covered the first-floor entrance of a keep.

***See 'small keep'*.**

Forecourt. The court formed by the walls of a barbican, or the barbican as a whole.

***See 'outer bailey'*.**

Foreland. The ground between the wall of a fortification and the moat.

Forework. (G. *vor*, fore). An outwork or fore building.

See 'barbican'.

Fork (Military). A polearm used to fend off siege ladders.

Fortalice. (Medieval Latin, fortalitia). A small fort or outwork of a fortification.

See 'castelletto', 'castlet', '*châtelet*'.

Fort à cavalier. ***See 'cavalier fort'.***

Fort à massif central. ***See 'cavalier fort'.***

Fort d'arrêt. (Fr. *fort stop*). A French isolated, detached, and advanced polygonal fort which was used to defend a passage, dating to the period between 1874 – 1914 CE.

See 'barrier de Fer'.

Fortaleza. (Sp. *strength*). A Spanish fort or fortress.

Fortezza. (It.). A stronghold, a fort, fortress, or castle.

Fortification. (L. *fortis*, strong; Fr. *facere*, to make).

(1) The act or art of fortifying a military position by means of defensive and/or offensive works.

(2) A work or structure, used as a military position; a fortified place or position.

Fortifications. Fortifications are usually divided into two types, permanent fortification and field fortification. There is also a third type known as a semipermanent fortification. The best analogues are the fortifications of the Romans;

(1) the *'marching camps'* were set up to protect the troops while marching through hostile territory, consisting of a surrounding ditch and rampart, with a palisade.

See '*marching camp*'.

(2) the *'legionary fort'* was planned out with defences, gates, roads, barracks, baths and toilets, granary, hospital, headquarters etc, built in stone.

(3) the *'castra'* or marching camps which had been upgraded with a proper stockade or stone walls, and added watch towers.

See '*castra*'.

Fortificazione alla moderna. (It. *modern fortification*). An Italian fort design dating from the 15th century CE based on the bastion system.

See 'arrow headed bastion'. 'bastion',' bastion system'.

Fortified bridge. A bridge provided with defences such as a tower(s) equipped with gates, machicolations, arrow slits and a portcullis.

The Monno Bridge in Monmouth, Wales, United Kingdom.

See ***'Brückenburg'*.**

Fortified church. From the earliest periods the use of a church as a place of refuge in times of strife was common, in some cases the church was provided with features of military architecture to increase its defensibility.

See 'church (fortified)', 'eglise donjon'.

Fortified dock, Fortified harbour.

(1) The fortified waterside part of a '*wasserburg*' or German water castle, the purpose of the castle was to control the traffic on the water way it fronted and to extract dues from the shipping.

(2) A dock or harbour which was provided with defences.

The Scaligero Castle of Sirmione on Lake Garda, Sirmione, Italy.

Fortified granary. A granary found in Morocco, consisting of storage cells where food products, valuable family assets and important documents are stored. The fortified granary at Agadir Inoumar, located around 70 kilometres (40 miles) from the city of Agadir in Morocco, is over 300 years old and occupies about half an acre, provided with a watch tower, water tank, a mosque and shelter for livestock and mounts, and was used as refuge when the need required.

Fortified group. A large defensive structure consisting of multiple elements from artillery and infantry emplacements, barracks, stores, all sunk into the ground and connected by tunnels.

See 'feste', 'fort palmé', 'Maginot line'.

Fortified line of resistance. Consisting of a line of '*supporting points*' providing mutual support with cross and flanking fire.

See '*support lines*'.

Fortified manor, fortified manor house. During the late 14th and the 15th centuries CE the necessity of fully defensible castles was on the decline, so that the number of fortified manors increased and developed into the undefended mansions of the Elizabethan and following periods. The plans of the fortified manors varied because of the addition of piece-meal defences to existing halls and manor houses. The most common design was of quadrangular plan, and were often surrounded by a wet moat. Also known as a '*defensible house*' or just a '*manor house*'.

Fortified posts. A system of posts raised by Charlemayne in the 7th and 8th centuries CE against the raids by the Saxons. Examples are in Eresburg and Karlstadt, Germany.

Fortified trading post. A trading post fortified against attack by local threats and enemy maritime powers (such as the Portuguese and the Dutch), and so assisting in the control of trade routes and access to precious commodities (like spices, silk and precious metals).

Fortin. (Fr.). A temporary earthwork fort used to protect a passage or as part of a circumvallation.

Fortlet. A small fort, which was used as a watch tower or as a signal station.

Fort palmé. (Fr. *fort webbed*). A kind of French fort used widely on the '*Maginot Line*', which had a huge irregular perimeter, defended by barbed wire and metal fencing, anti-tank ditches, provided with underground facilities like barracks, stores, observational posts, machine gun casemates and artillery bunkers, all connected by tunnels.

See '*Maginot Line*'.

Fortress. (O.Fr. *fortresse*). A strong permanent fortification which may include a town or attached to a town.

Forward operating base (FOB). A local secure military operation facility that supports tactical and strategic goals, and allows for fast reaction times to developing combat situations. May contain logistical facilities, mechanical workshops, hospital and air base. Normally supported by a main base of operations.

See 'main operating base'.

Foss, Fossa, Fossae, Fossata, Fossatum, Fosse. (L. *folio*, dig).

(1) A ditch, moat, canal or pit, which was used to impede the advance of an enemy force.

See *'ditch'.*

(2) The ditch or a set of ditches which surrounded a Roman fortification, the excavated soil was used to construct the '*vallum*' or '*rampart*'.

See 'agger', 'ditch', 'dyke'.

Fossatum Africae. (L. *African ditch*). The fortified southern border of the Roman Empire in northern Africa, consisting of ditches and ramparts, numerous forts and small watchtowers, built in 122 CE. Also known as the '*African limes*' or '*African ditch*'.

See 'Antonine wall', 'foss', 'Hadrian's wall', 'limes', 'limes Raetian'.

Fossa fastigata. (L.). A type of Roman ditch described by Hyginus (1st century CE) that the ditch was '*V*' shaped in cross section. These ditches could be provided with a narrow channel in the bottom.

See 'foss'.

Fossa punica. (L.). A type of Roman ditch described by Hyginus (1st century CE) that had a much steeper outer scarp, and was used as the outermost ditch.

See 'foss'.

Fossatum Africae. The African limes. **See 'limes'.**

Fossé diamant. (Fr. *diamond ditch*). An angular concrete ditch usually in front of an embrasure or embrasures of a casemate, dating from 1874 CE, built as part of the fortifications of the '*Séré de Rivières forts*' of France and later on the Maginot line. Such a ditch could measure between 1 to 5 metres (3 to 16 feet) wide and 3 to 4 metres deep (10 – 13 feet). Used not just as an obstacle, the ditches also were used to collect debris that would otherwise obstruct the embrasures, to collect the casings from machine guns thus conceal them and the position of the embrasures, sometimes the ditch was defended or housed the emergency exits.

See 'ditch', 'foss', 'Maginot Line', 'séré de rivières system'.

Four-leaf keep. A keep consisting of four round towers which mutually intersected, dating from the 12th century CE.

See *'quatrefoil keep'.*

Fougasse. An improvised mortar consisting of a hole cut in the rock or dug in the ground at 45 degrees aimed at a beach, filled with an explosive charge and then rocks, used as anti- personnel weapon. An example is the Fourasse of Malta, by the Knights of St John in the 18th century CE.

Fox hole. A small hole dug into the ground for protection from enemy fire. All so known as *'fighting position'* (US Army), *'fighting hole'* (US Marines), *'gun pit'* (Australian Army), *'fighting pit'* (New Zealand Army), *'rifle pit'* (US Civil War by both Union and Confederate Armies.

Fraise. (Fr.). A horizontal or inclining palisade which was placed on the outward slope of a rampart or the berm of an earthwork, to prevent the work from being taken by surprise and to impede an enemy's advance.

Freccia. A detached lunette attached by a long corridor or caponier to the main work.

See *'arrow', 'lunette'.*

Frizzy horse. A movable obstacle composed of crossed wooden stakes, joined by a cross beam, which could have barbed wire strung from the various parts, used to block roads, passages or gaps in the barbed wire network.

See *'cheval de fries'.*

Frontière de fer. *See 'pré carré'.*

Fronting. The orientation of an earthwork towards the enemy's attack.

Front ditch. The ditch on the outside of the parapet provides an additional obstacle to the attack of an enemy. Also known as an *'exterior ditch'.*

See *'ditch'.*

Frontiera Nord. (It. *northern frontier*). The fortified northern border of Italy with Switzerland, consisting of 72km (45 miles) of trenches, artillery positions and barracks, building starting in 1899 CE and was strengthened during WWI, built to defend against possible invasion into the Po Valley by either France or Germany (Austria-Hungary) via Switzerland but was never used.

See 'alpine wall'.

Front line. The *'fire trench'* or the first line of defence, and from where troops would attack from.

See *'support trench'.*

Front of a bastion. The distance which lies between two adjacent bastions.

Front Todt. The semi-circular stepped concrete canopy constructed above the embrasure of some types of heavy coastal artillery bunkers in the Atlantic wall.

See 'Atlantic wall'.

Longues-sur-Mer battery, Normandy, France.

Full bastion. When the terrain within a bastion is at the same height as the rampart with only the parapet rising up to protect the soldiers behind, the bastion is said to be full.

See *'bastion'.*

Fukugôshiki tenshu. See *'teiritsushiki tenshu'.*

Furrow of glacis. The line where two glacis meet at the *'re-entrant angle'* of the *'place of arms'* and the *'covert way'*.

See 'glacis'.

Futaebei. (J.). A type of castle wall constructed on two parallel walls with the gap between the walls is filled with earth, sand and stone, to counter the use of firearms. Also known as *'taikobei'*.

FW3. The British Directorate of Fortifications and Works was set up in 1940 CE, and task with providing designs for pillboxes for infantry defence with rifles and machine guns, along with pillbox like structures for anti-aircraft positions, observation posts, search light positions, for defending landing sites, river and canal access, and air fields.

See 'allan williams turret', 'cantilevered pillbox', 'dover quad', 'eared pillbox, 'eastern command variants', 'essex lozenge pillbox', 'field gun emplacements', 'FW3', 'Lincolnshire three-bay', 'lozenge pillbox' 'northumberland 'D' Type', 'pentagonal pillbox', 'pillbox', 'pill box 22,

23, 24, 25, 26, 27, 28', 'ROF Type', 'section post/seagull trench', suffolk square', 'vickers machine gun emplacement'.

G.

Gabion. (It. *gabbione*, big cage). A wickerwork basket which was filled with earth and used to build field works, such as revetments and parapets, also used by sappers as cover from musket shot, as they advanced their trench the gabion was rolled before them.

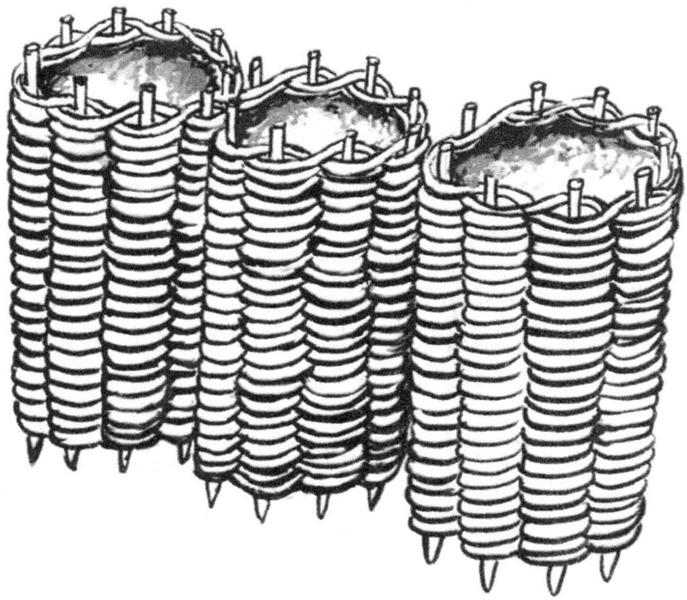

See 'bastion', 'bottom', 'corbeil', 'hexo barrier'.

Gabionade. A field work which was constructed using gabions.

Gabion revetment. A revetment using gabions to stabilise the embankment or rampart, usually topped with sod or sandbags.

See *'gabion', 'revetment'.*

Gads. Ties made from willow saplings used to tie willow saplings into bundles to make *'fascines'*.

See *'fascines',*

Gaffs. The pair of wooden lifting arms that protrude from the slots (Fr. *rainures*, groves) in the gatehouse of a castle, which were connected to the wooden deck of a bascule bridge by chains.

See 'bascule bridge', 'bridge pit', 'drawbridge', 'gatehouse', 'rainures'.

Gakejiro. (J.). Clifftop castle.
Galleria cannoniera. (It. *gun gallery*). A rock cut battery consisting of one or two artillery positions, connected to other works by tunnels. Also known as a '*cave battery*'.

See 'alpine wall', 'frontiera nord'.

Gallery. A passage connecting the inner and exterior works of a fortification, which was either built of masonry or an underground passage.

See 'caponiere', 'traverse (2)'.

Gallery of the counterscarp. A defensive position situated in the counterscarp or the outer side of the ditch of a fortification, which was used to flank the ditch, also known as a '*counterscarp gallery*'.
Galopin turret. A type of French eclipsing armoured turret from the 1890's CE, provided with artillery, ventilation system and a lift for ammunition.

See 'eclipsing turret', 'tourelle galopin', 'turret'.

Ganerbenburg. (G. *inherited castle*). A German castle under multiple ownership by different parts of the same family. Separate homes were set up in different parts of the castle.

See 'hauser'.

Gan ge ban. (Ch. *conflict plank*). **See '*cha ban*'.**
Gangizake. (J.). A wide set of stone steps ('*ishidan*') leading straight up a slope of an earthen embankment, allowing a large number of defenders to ascend the steps at once. Also known as a '*mushabashiri*'.

See '*kasanezaka*'.

Garh. (Urdu). Fortress.
Gariton. (Sp.). A Spanish castle's sentry post, typical of the 15th century CE.

See '*bartizan*', '*echaugette*', '*guerite*'.

Garrison. (OFr. *provide defence*). The soldiers guarding or defending a fortification. The number of soldiers in a garrison is hard to determine, in times of peace the number of soldiers would be small, enough to fill the day and night watches, enough to man the gates and provide patrols of the surrounding area. The soldiers could be raised by a levy on the surrounding lands owing feudal duty or paid a wage or stipend. The numbers of soldiers could increase tenfold in times of war.
Gate. (Old English *gæt, geat*). A gate is used to block access through a gateway, normally made of timber, sometimes reinforced with iron, and could be provided with a '*wicket*' for access when the main gates

are closed. Gates could be secured by locks or cross beams held in place by '*put log holes*' on the sides of the gateway.

See '*gateway*', 'zangentor'.

Gatehouse. The structure which was used to guard the principal entrance of a fortification. A gatehouse usually consisted of the following. a single archway or passage; a stout gate or drawbridge; and the option of a portcullis either at the front of the entrance, or two portcullises one at either end. The vaulted ceiling of the entryway could be provided with murder holes, the entrance was generally flanked by projecting towers which were pierced by arrow slits, and the upper parts of the towers terminated with machicolations and battlements.

See '*barbican*', '*forebuilding*', '*great gatehouse*', '*kammertor*', '*masugata mon*', '*mon*'.

Gate tower. A tower which guarded a gateway, usually two such towers were used to flank the gateway, one on either side.

See 'gatehouse', 'great gatehouse'.

Gateways (Baffles and Bastions). Fortification gateway came in many forms including the following types:

- Simple baffle, consisting of two overlapping walls with a gateway between them making it necessary to make a right turn on entry;
- Bent axis (Claviculum), where a 'L' shaped wall meets a straight wall making it necessary to make a left hand turn on entry;
- Offset, where a '*L*' meets a '*T*' shaped was making it necessary to make a left then a right turn on entry;
- Oblique, a gateway between two protrusions from the walls set at an angle (45 degrees) causing an entrant to change course;
- Crab claw, a gateway where one wall terminates in a 'C' shape and the opposite wall is straight and is sited between the arms of the 'C', causing the entrant to change directions at least 4 times;
- Serpentine, a complicated gateway, with a '*C*' shaped side, offset with another wall with a '*C*' shaped terminal, which would require an entrant to change direction at least 6 times;
- Labyrinthine, a complicated gateway with a number of twist, turns and dead ends, allowing the enemy to be trapped;
- Screened (Titulum), A break in a wall, with a separate wall in front of the gateway, parallel to the main walls, may also have a parallel wall behind the gateway.
- Insert / Offset, a '*L*' shaped projection from a wall end, either to front or rear of the gateway;

- Bastioned (Barbican), a direct entry gate with a bastion protecting the gate way on one side or both;
- Chambered, a '*C*' shaped chamber on either side of the gate of a direct entry gateway. May include guard rooms on either side of a gateway, allowing guards to be stationed in close proximity to the gate.

Gaztelu. A hillfort dating from the Iron Age located in the Basque country (northern Spain).

See 'hillfort', 'oppida'.

Gazion. (Fr. *gazon*, grass). A piece of turf which was cut into a wedge and used with others to line the outside of earthworks such as parapets and the traverses of galleries, to hold them in place.

See 'cespites', 'revetment'.

Gegenburg. (G. *counter castle*). A counter castle built by besiegers to starve out or neutralise a castle.

See ***'counter castle'***, ***'malvoisin'***.

Genoese towers. The Genoese erected towers or refurbished old towers to protect themselves and their business wherever they went, acting as sentry posts, refuge towers. Early towers were square in cross section, later they were round in cross section.

Examples are;

(1) Galata Tower (Istanbul) built in 1348 CE and is 67 metres (219 feet) high.

(2) Tour de Capitello, south west coast of Corsica, built in 1552 CE and is 11 metres (36 feet) high.

Tour de Capitello.

Geschützstand. (G. *Gun emplacement*).

Geschützstellung. (G. defensive position). Gun emplacement.

Giri-durga. (Sanskrit). A hill fort. Also known as *'parvata durga'*.

See 'durga'.

Giri-parshva-durga. (Sanskrit). The fortifications and civil structures that extend down the hillside from the *'giri-durga'*.

See 'durga'.

G-Kazemat. A large type of Dutch reinforced concrete casemate, a 1930's CE part of the pre-WWII defensive program, defended by a heavy machine gun or field gun, with an armoured door on the flank to the embrasure. Sited at road crossings, dykes, river fords where they could provide direct or flanking fire. The *'G-Kazemat'* were provided with supportive smaller bunkers to protect its flanks or by obstacles like anti-tank ditches.

See 'casemate'.

Glacis. (L. *glucies*, ice).

(1) The area outside the ditch which was scarped into a gentle slope running downwards from the covered way towards the open country, which was kept deliberately free of any form of cover. The glacis brought an approaching assailing force into clear view from the parapet of a fortification under attack.

See 'declivity'.

(2) The masonry sloped scarp of a curtain wall, a design which was developed to offset the effect of artillery fire.

See 'talus'.

Glacis parapet. The parapet directly behind the glacis.

Glacis rampart. *See 'dump rampart', 'rampart'.*

Glasmine 43. An anti-personnel mine which was hard to detect because it was made of glass, used by Germany in WWII.

See *'mine'.*

Gomonban nagaya. (J.). A castle guardhouse.

Gonnes. *See 'cannons'.*

Gopur. (Sanskrit). The main gatehouse of a fort which is two or three stories high, with a two leaved gate studded with *'parighas'* (iron spikes).

Gorge.

(1) The interior side or entrance of a bastion or other outwork, which is usually the interval between the two flanks of the work.

(2) The rear or any part of a work which is next to the body of the place at the counterscarp of the ditch where there is no rampart.

Gord. (Slavonic). A fortified settlement built on a strategic site like a hilltop, river bank, island in a lake or a peninsula, consisting of buildings surrounded by an earthen rampart with a wooden palisade, dating from between the 6th to 12th centuries CE.

See *'gorod'.*

Gorod. Russian fortified settlement.

Götavirke, Geatish Dyke. A set of parallel defensive walls running north/south, from Västra Husby to Hylinge in Östergötland, Sweden. Dated to around the year 800 CE, 3.5 km (about 2 miles) long, is a 6 to 10 m (20 to 32 feet) wide and 0.6-2.5 m (2 to 8 feet) high, supplied with a palisade and moat on the Eastern side, to defend against attacks from the Baltic Sea.

See '*dyke*'.

Goten. (J.). A castle's palace.

Gothic line. One of the defensive lines built by Organisat Todt in Italy to slow the Allied advance in WWII, situated along the summits Apennine Mountains, 320 km (200 miles) long with minefields, bunkers, machine gun nests, anti-tank positions, barbed wire and anti-tank ditches. The name was changed to the 'Green Line" (G. Grüne Linie) in 1944 CE.

See '*Gustav, Hitler and Reinhart lines*'.

Goulette à Grendage. A pipe or duct within the concrete wall of an emplacement to drop grenades on enemy troops, used in the Maginot Line.

See 'créneau de pied', 'Maginot Line'.

Grade. The original ground level, where the '*ditch*' is below grade and the '*parapet*' above grade.

See '*ditch*'.

Gran bugue. A Spanish fortress dating from the 15th century CE, which was built in the shape of a great battle ship. The enceinte provided with mural towers formed the bow and the stern while the keep formed the bridge, normally situated on hill tops. The term literally means '*great ship*'.

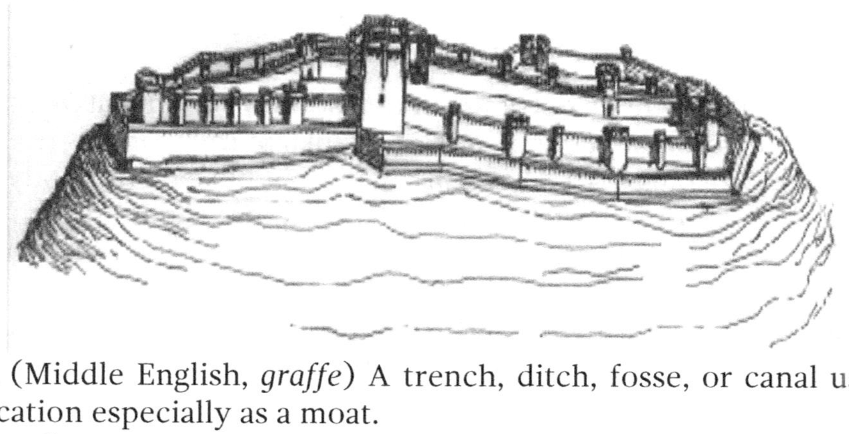

Graff. (Middle English, *graffe*) A trench, ditch, fosse, or canal used in fortification especially as a moat.

Grazing fire. Weapons fire that skims off a surface.

See *'fire'*.

Greater fieldwork. An earthwork constructed with a front or exterior ditch. Also known *as a 'major fieldwork'*.

See *'ditch'*.

Great gatehouse. The gateway of a castle was always a major weak spot in the defences, which later developed into the strongest part of a fortification's perimeter. The development in gatehouse design known as the '*great gatehouse*' has been compared in strength to the keeps of earlier castles. Using the same principles as in the gatehouse, the great gatehouse was provided with a pair of large flanking towers projecting forward from the line of the curtain wall, while two smaller flanking towers projected into the inner ward.

The gatehouse was provided with a number of defences such as; machicolations, slits and battlements, the passageway itself was defended by a portcullis at each end and murder holes in the vaulted roof above. The chambers above the passageway were used to house the machinery for raising and lowering the portcullises and the drawbridge if one was provided, the rest of the remaining area was used as accommodation for the garrison or sometimes as residential apartments for the lord or castellan of the castle.

See *'concentric castle', 'Edwardian castle', 'keep gatehouse', 'ring of iron'*.

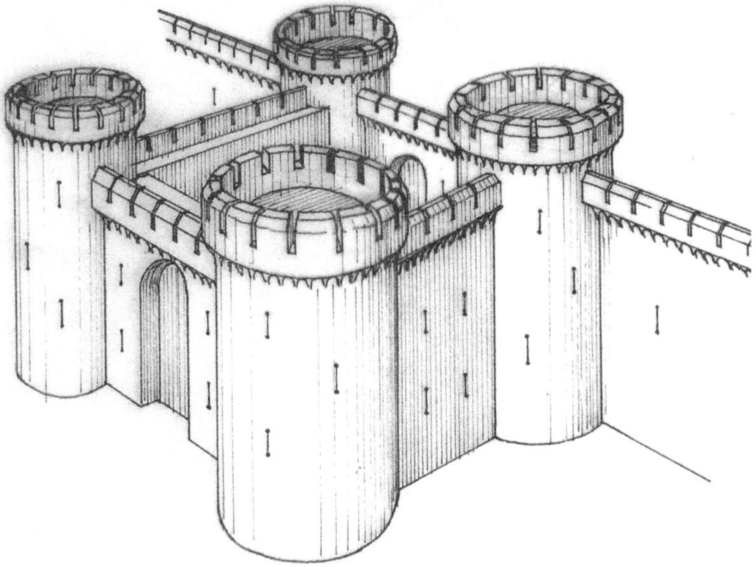

Great tower. See *'keep'*.

Great wall of China. The Great Wall of China extends over 6350 kilometres (3945 miles), and consists of multiple sections from Shanhai pass in Bohai Gulf on the coast to Mongolia in the north and North Korea in the south. Constructed by various Chinese dynasties, starting in 220 BCE by the Qin Dynasty.
The most recognised sections The Great Wall was built during the Ming Dynasty (1368-1644 CE) and is most visited. It consists of crenelated walls on each side of the wall top, with watch towers, major towers and gates spread along its length, built in stone, brick, timber, and in the desert they used clay and reeds. When possible, they incorporated natural barriers like mountains and rivers.

Katiebell © 2024

Great wall of Gorgan. (Sassanian). A defensive wall of the Sassanian Empire against the Huns, built in the 5th and 6th centuries CE, running from the Caspian Sea and the mountains of in north east Iran, 195 kilometres (121 miles) long, 6-10 metres (20-33 feet) wide, A wall provided with 40 forts built at varying intervals. Built of a range of materials including; mud brick, fired brick, gypsum, and mortar.

Great Wall of India (Kumbhalgarh). (lit. "*Kumbhal fort*").
A hill top fort in Rajasthan, India, dating from the 15th century CE, spanning 38 kilometers (24 miles), protected by crenelated walls and bastions, punctuated by eight gates.

Great wall of Korea. (Korean, *Cheolli Jangseong*). Two separate defences built between Korea and China. The first was built in the 7th century CE (500 km (310 miles)) raised against the Tang dynasty China consisting of a network of military garrisons from Buyeoseong and Bohai Bay.The second was built in the 11th century CE, about half as long as the first (to be verified), raised against the Khitan and the Jurchen, consisting of stone walls from the cities of Uiju to Hwaju.

Great wall of Qi. (Ch. *Qí Chángchéng*). About 500 BCE the Qi Dynasty of China built a wall to protect its borders from attack of the Jin and Yue states, running 595 kilometres (370 miles) from the Shandong Province to the Yellow Sea, made mostly of stone and rammed earth.

See '*Great wall of China*'

Great wall of Sindh (Ranikot Fort). Talpur-era (17th century CE) fort near Sann in Jamshoro District, Sindh, Pakistan, with walls 31 km (19 miles) long. The northern perimeter is a natural barrier of steep hills with the other three sides protected by fort walls. The fort's crenelated walls are provided with several bastions and four gates. And there is a small fort inside the perimeter known as the '*Miri Fort*' which is about 3 km from the Sann gate.

Great Zimbabwe. An irregularly shaped dry stone enclosure with various gates and courtyards, walls are up to 10.5 metres (34.5 feet) high and between 3 and 5 metres (10 and 16.5 feet) thick, perimeter measures about 300 m (984 feet), probably dating between the 6th and 10th century CE. Located about 250 km (155 miles) south of Harare (Zimbabwe).

Grebbe line. (Du. *grebbelinie*). A Dutch defence line of the Dutch Waterline, based on flooding and defence works, established in 1745 CE to protect the Netherlands from invasion, which was re-fortified in the 1930's CE, still relying on flooding but along with trench works and a variety of casemates.

See 'B-Kazemat', 'G-Kazemat', 'pyramid', 'Kanon Kazemat 6 Veld', 'S-3 Kazemat', 'S-4 Kazemat'.

Greek fire. First used by the Byzantines in ship to ship battles, i.e. against the Rus in the 10th century CE, involving squirting a flammable liquid out of nozzles at the front of their ships, acting like a *'flame thrower'*. Various ingredients have been proposed, including; petroleum, resin, naphtha, crude oil, pitch and quick lime. At Montreuil-Belley (France), Geoffrey V of Anjou, in 1151 CE, used Greek fire flung by a mangonel, burning three houses. Later, the Arabs used Greek fire to defend Acre against members of the Third Crusade (1189–1192 CE) using siege towers.

Green Line. (G. *Grüne linie*). ***See 'Gothic line'.***

Green zone. A fortified zone in a city which houses government buildings and diplomatic missions (i.e. Iraqi capital Baghdad, 2021 CE).

Grillage. A mesh of timbers (usually cedar because of its resistance to rot), often in layers, used as a foundation of a fort on boggy terrain.

Grille. An iron grating defending a window in a wall or in a door. Examples can be found at: Castle of Coucy, in Picardy, France; and at Carcassonne, Occitania, France.

See *'manidwar', 'speculartorium'.*

Grille Belge. ***See 'Belgian gate'.***

Grody. A Slavic fortress.

Groepshelter – type P.

See 'pyramid', 'casemate'.

Groepsshiuiplaats. ***See 'pyramid'.***

Groupe fortifié. ***See 'feste', 'fortified group'.***

Gruson turret. An armoured turret used in forts of the 19th century CE, designed by a German company (Gruson), used at several forts in the Weser estuary (Germany) for coastal defences, and at the Italian naval base at La Spezia.

Guardiola. (It. *guard room*). ***See 'echauguette', 'guérite'.***

Guard room. A room situated at the main gate to a fortification to house the guards guarding the gate.

Guardinghi. Italian turreted buildings used as watch towers by the Langobards.

Guard wall. See *'curtain wall'.*

Guelph merlon. A merlon provided with a *'V'* shaped apex, common around the 14th century CE.

See *'merlon'*.

Guerite. (Old Fr. *garite, guarite,* watchtower). A sentry box situated on a bastion or curtain; also known as an *echaugette.*

See *'echaugette'*, *'gariton'*.

Guha-durga. (Sanskrit). A fort sited in a valley surrounded by hills, where the outposts and signal towers are located.

See 'durga'.

Guichet. (Old Norse *vik*, Middle Dutch *wiket, winket*, wicket). A small opening in a wall, or a small door in the gate of a fortification.

See 'klinkett', 'postern', 'wicket'.

Gulaigorod. (Russian, *wandering town*). A field work used in 15th century CE Russia. Large shields were provided with wheels and loop holes. When they camped the Gulaigorod were drawn into a circle, thus presenting a fortified perimeter.

See 'carrago'.

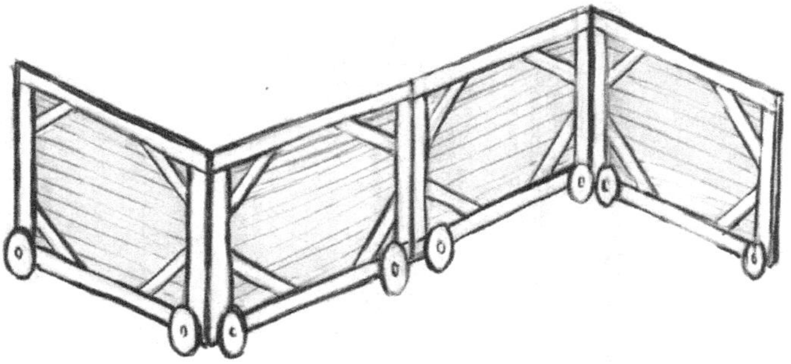

Gun house. An armoured shield for a gun which had a roof, front, sides but not covering the rear on a barbette mount.

Gun loop, Gun port. An aperture through the wall of a fortification for firing guns at an enemy. Their purpose was the same as an arrow loop, which was to provide covering and flanking fire. They took many forms but the main types were: the blind gun port, consisting of a simple circular opening; the circular hole with a vertical sighting slit above the opening; also other types had different variations of the sighting slit.

Gun ports were built into the walls and battlements where the guns could be directed in relative safety against attackers. The placement of gun ports was usually at the weakest part of the fortification, which was generally the gate.

See *'blind gun loop', 'cross and orb gun loop', 'keyhole gun port', 'loop', 'port'.*

Gunports. Gun ports came in a range of shapes and sizes, such as:

- Circular - Wall towers Topakapi Palace, Istanbul;
- Cruciform – Ragland Castle, The great tower , 15th century CE;
- Dumbell – Threave Castle, Galloway, 1447 CE;
- Horizontal (widemouthed) – Holyroodhouse and Falkland in about 1500 CE;
- Keyhole – Cooling Castle, Kent, on the inner gateway 1381 CE;
- Rectangular – Devil point blockhouse, Plymouth, 1537-1539 CE.

Gun pit. An entrenched artillery platform in the form of a '*demilune*' or an '*epaulement*'.

See '*dummy gun pit*', '*fox hole*', '*tobruk*'.

Gun platform. An unroofed platform which was specially built to provide an area for the cannons of a fortification. The gun platform was a much better solution to artillery defence than gun ports because the noise and fumes of gun fire in the confined space made it very uncomfortable to fire more than a few rounds.

See '*cavalier*'.

Gun powder, gunpowder. A mixture of sulfur, carbon, and potassium nitrate. Introduced to Europe in 1320 CE from China. Used in cannons, mortars, petards, muskets etc., leading to fundamental change in fortification design. Gone are the lofty towers, and in are the low earthen embankments of defences in depth with reduction in dead ground.

Gunpowder mine.

(1) A countermeasure to a mine (invention credited to John Vrauo) used to blow up the mine placed in the counter mine (like the Turkish mine at Belgrade in 1433 CE).
(2) The countermeasure could be used to create a breach instead of the usual burning out the mine supports.

Gun ramp. A ramp used to move a cannon into a firing position on a gun platform.

Gun towers. Towers provided with cannon. See Cowdry's painting of the embarkation of Henry VII to the Cloth of Gold tournament in France (1520 CE) for two different types of '*gun tower*'.

See '*Martello towers*'.

Gurry. (H. *garh*, hill fort). A fort which was constructed of mud, and the walls were flanked by towers, and sometimes ditches surrounded the fort, used in India.

Gürtelfestung. (G. *belt fortress*). A fortress surrounded by several detached forts, which provide protection to the core fortress. Also known as a '*camp fortress*'.

Gustav line. One of the German defensive lines built by Organisat Todt in Italy to slow the Allied advance in WWII, situated behind the Garigliano and Rapido Rivers with the monastery Mount Cassino at its centre (150 km, 93 miles long), made with pre-fabricated turrets and cupulas. Delaying the Allied advance in the winter 1943 to 44 CE.

See 'Gothic, Hitler and Reinhart lines'.

Gusuku. (Okinawan, *gushiku*). Castles of Okinawa Ryukyu Islands which feature stone walls which usually follow the contours of the land, with one or more baileys and one or more entrances. Dating from the 15th century CE, they were built extensively of stone, lacking the timber superstructures which were typical of the mainland castles. Later in the period the '*gusuku*' were modified to have gun ports to protect the gates.

Guthrie rolling bridge. An unhinged retractable bridge over a moat, used on 18th and 19th century CE British forts.

See 'drawbridge'.

H.

Habicht turret. A German WWII rotating turret made of concrete because steel was in short supply, only one was built near Calais (France) as part of the '*Atlantic Wall*'.

Hadrian's Wall. A wall similar to the Germanic Limes or the Antonine Wall, originally constructed of earthen ramparts, later replaced in stone, with eighty milecastles, numerous observation towers and seventeen larger forts, 117.5 km (73 miles) long, built under the direction of Hadrian (AD 117-38 CE) which formed the Roman frontier across England from Wallsend, on the Tyne, to Bowness on the Solway Firth.

See 'Antonine wall', 'limes', 'limes Germanicas', 'limes Raetian'.

Hagasane. (J.). Long overlapping iron plates attached by nails to cover a '*tenshu's*' walls, overlapping from top to bottom of plates, including window posts and latticework (an example is the north side of Fukuyama castle, Hiroshima prefecture, Japan, dating to Early Meiji period (1868 to 1912 CE)).

See '*teppan*'.

Hairpin. A British WWII type of anti-tank obstacle made from bent railway line or girder embedded in concrete blocks. Extant examples can be found in Narborough and Narford, Norfolk, England.

See 'obstacle'.

Hakobori. (J.). Box shaped trench.

See 'bori', '*kenukibori*', '*shojibori*', '*tatebori*', '*unebori*', '*unejotatebori*', '*yagenbori*'.

Hako hazama. (J. '*hazama*', to fire a gun or loose an arrow). A square shaped loophole.

See 'ishisama', 'sama', 'shinogi hazama', '*taihosama*', '*tepposama*', '*yasama*', '*zama*'.

Half bastion. The half of a bastion made when bisected at the capital, consisting of one flank and one face.

See '*bastion*', '*demi-bastion*'.

Half merlon. The merlon at the end of a parapet that is half the size of the other merlons. **See '*merlon*'.**

Half moon. See '*demilune*', '*moon*', '*ravelin*'.

Half moon barbican. An outer defence in the form of a semi-circular barbican.

See '*barbican*'.

Half moon battery. A battery in the shape of half a moon. Examples are at Carlisle castle and Edinburgh castle, UK.

Hall keep. Originally a type of rectangular keep, which later developed into a round design with more than one storey, including a main hall and apartments.

See '*keep*'.

Halsgraben, (G. *neck ditch*). A dry ditch defending an approach to a hill castle, or cut a peninsula leading to a lowland castle. Also known as a '*throat ditch*'.

Handgun loop. A gun loop which was especially designed for the use of hand guns.

See '*gun loop*'.

Han cheng. (Ch. *guarding wall*). A small auxiliary wall built at the base of the outer side of a main wall, some 2 metres (3 feet 3 inches) high, designed to reinforce the main wall against damage from the frequent river floods.

See '*cheng*'.

One of the gates of Taizhou city wall, China.

Hane-age-mon. (J. *wing raised gate*). A type of gate used in the defence of a Japanese castle which consisted of wooden or bamboo grid frame hinged to the top of wooden gate from, swinging outwards, and needed propping to hold the gate open. Examples can be seen at Asuke Castle in Aichi Prefecture.

See '*drawbridge*', '*hikibashi*', '*porticulis*', 'tapecu', '*yett*'.

Hangburg. (G. *hillside castles*). A German castle built on the side of a hill below the summit, dating from the second half of the 11th century

CE. The side facing the summit was usually protected by a shield wall (*'schildmauer'*) or a *'bergfried'*, sometimes both.

See *'castle'*.

Ehrenfels Castle, above the Rhine river valley, Germany.

Hanging turret. See *'bartizan'*.

Hanmasugata. (J.). The square shaped defensive enclosure situated at the main entrance of a castle (*'koguchi'*), where one half of the enclosure is situated inside the castle perimeter, while the other half projects out of the enclosure.

See *'barbican'*, *'masugata'*, *'masugatamon koguchi'*, *'masugata mon'*, *'sotomasugata'*, *'uchimasugata'*.

Hao. (Ch. *moat*). A defence structure dug in front and around a fortification, which could also serve with flood control. Also known as a *'hu cheng he'* (Ch. *city protecting the river*).

Hardened aircraft shelter (HAS). A reinforced hanger used to house and protect military aircraft from attack, usually used for fighter aircraft. Also known as a *'protective aircraft shelter'* (PAS).

Hashidai. (J.). An earthen bridge with walls along both sides used to connect a *'daitenshu'* to a *'kotenshu'*.

See ***'reritsushiki tenshu'*, *'renketsushiki tenshu'*.**

Hauptmauer. *(G.)*. The main wall of a castle.

Hauser. (G. *house*). The part of a German *'Ganerbenburg'*, which was occupied by a branch of the family to whom the castle belonged.

See *'ganerbenburg'*.

Hawsers. (Middle French *hauci(er)*, to hoist). Ropes strung between metal rings attached to buildings to unseat German reconnaissance motorcyclists landed by gliders as part of an invasion. Examples of the metal rings can be found in Church Street, Ampthill, Bedfordshire in the UK.

Haxo Casemate. A casemate, or series of casemates, formed in the parapet or freestanding on the terreplein, used to house cannon. Designed by Baron Haxo (François Nicolas Benoît) a 19th century CE French Army general and military engineer.

Head log. A wooden log on top of a parapet with gap underneath provides a sniper or infantry a spot to fire from without excessive exposure to enemy fire.

See *'skids'*.

Headquarters. A building site or out in the field where the commanding officers and their staff were housed either temporally or long term.

Hedgehog. See *'Czech hedgehog'*.

Hedgehog defence. The use of multiple strong points which are mutually supportive thus provided *'defence in depth'* and able to slow the advance of an enemy and then counter attack, providing enfilade fire, thus each *'hedgehog'* has to be overcome before an enemy can move forward.

***See 'Army anti-tank island', 'defence in depth',* 'panzerstutzpunkt', *'stop lines'*.**

Hei. (J.). Defensive fences used to partition *'kuruwa'* or on the top of stone or earthen walls.

Heinrich. A German WWII field shelter for three to six men, made with earth and corrugated iron.

Heishitaji. (J.). The earth used in a castle's earthen walls, which is reinforced with pieces of bamboo laid horizontally and vertically, with the tops of the vertical bamboo poking out the top sharpened creating anti-personal defence.

Helepolis. (Gr.). A siege tower.

See *'siege'*.

Hemmbalk. (G. *hemm bar*). A German WWII beach obstacle consisting of timber or concrete beam sloping inland at 30 to 40 degrees, placed below the tide line in alternating rows to strand landing craft. The hemmbalk could have metal blades (*stahlmesser*) or anti-tank mines (*tellermine*) to add to the destructive effect.

See 'admiralty scaffolding', 'Atlantic Wall', 'belgian gate', 'hemmkurven', 'Rommel's asparagus'.

Hemmkurven. (G. *hemm curves*). A German WWII anti-tank and landing obstacle, made from five steel bars welded together, developed in 1937 by defunct by 1943 CE, used on beaches where piles could not be driven into the beach. Also known as a '*strand bar*'.

See 'beach obstacles'.

Herclian Wall. A double wall extending from the north-western angle of the enclosure around the Palace of Blachernae to the Golden Horn, connected by transverse walls, the inner wall was constructed in 627 CE by Emperor Heraclius, and the outer wall was added by Emperor Leo V (early 9th century CE).

See 'Constantian Wall', 'Theodosian Land walls'.

Herisson. A long iron spiked wooden beam set on a pivot. When rotated, it always presented a front of spikes, its purpose was to block passages and approaches.

Herse. (Anglo-Norman French *herce*, harrow frame). A door stuck with protruding iron spikes. The door was hinged at the top and suspended in the open position by a rope. This rope was cut or released to affect a surprise blockage of the gateway or passage where it was situated.

Hersillon. A beam stuck full of protruding iron spikes which was thrown into a breach to render it impassable.

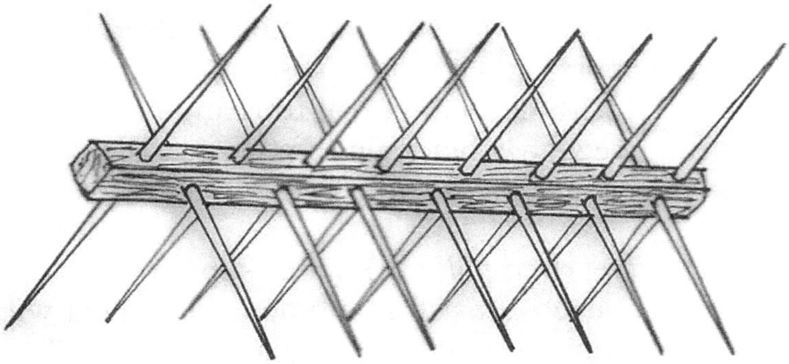

See 'abatis', 'barbed wire', 'chevaux de frise', 'entanglements', 'zaqaziq'.

Hexamilion wall. (Gr. *six-mile wall*). A defensive wall built across the Isthmus of Corinth, from the Gulf of Corinth to the Saronic Gulf (8km, 5 miles long), by Emperor Theodosius II, blocking the land route from the Peloponnese peninsula from mainland Greece, to protect against the Visigoths. Built between 408 and 450 CE, provided with curtain walls and square towers.

See *'Anastasian wall', 'Antonine wall', 'Çakmak Line', 'Hadrian's wall', 'Theodosian Land walls'.*

Hexco barrier, bastion. a modern (1989 CE) gabion (a collapsible wire mesh with a heavy duty fabric lining) which can be filled with local soil/sand/rocks (usually by front end loader), to provide form defences offering protection from explosions, blast and small arms fire (i.e. Iraq, Afghanistan Wars).

See '*gabion*'.

High wire entanglement. A tall wire entanglement on wood pickets 10 cm (4 inches) in diameter and 1.8 m (6 feet long) sunk into the ground about 0.6 m (2 feet), with barbed wire at tops and bottoms, and strands of barbed wire run irregularly between the two strands, set out in rows.

See '*barbed wire*', '*entanglement*', '*low wire entanglements*', '*spider wire entanglements*'.

Hikaehashira. (J. *supporting pillar*). Timber supports at right angles to the inside of a '*dobei*', each consisting of one upright post with mortises for two horizontal beams running into the back of a '*dobei*'.

See '*dobei*'.

Hikaegui. (J.). An upright wooden pole supporting a castle's wall.

Hikibashi. (J. *pull bridge*). A wooden bridge hinged at the bottom and moved by ropes or chains at the top (when closed) to secure access to a Japanese castle. The wooden bridge could be dismantled and moved if required.

See '*drawbridge*', '*hane age mon*', '*porticulis*', '*tapecu*','*yett*'.

Hikuhashi. (J.). A wheeled bridge that could be retracted to prevent an enemy from entering a castle. Also known as '*sorobanbashi*'.

Hilani edifices. The turreted defences of the Hittites (i.e. the fortress of Sinjerli (10th to 8th centuries BCE).

Hill fort. A fort which normally covered an entire hill top, sometimes as much as an area of several acres, and was enclosed by either a dry-stone wall or earthen ramparts. A protective ditch was on either the inside or outside of the surrounding wall.

See '*multivallate hillfort*', '*vitrified hill fort*'.

Hillside castle. See '*hangburg*'.

Hilsea Lines. A defensive line built to protect the northern approaches to Portsea Island, off the southern coast of England, thus protecting the naval base and city of Portsmouth. Built in 1756 and 1757 CE, consisting of a ditch and rampart, provided with sluices at either end. Between 1858 and 1871 CE the lines were modified with bastioned trace and fortified bridges for rail and road access. For more information see the fortifications of Portsmouth.

Hindenburg Line (Siegfriedstellung, Siegfried Position). German defence lines built well behind the main German front lines in 1916 - 1917 CE, 145 km (99 miles) long, from Tilloy-les-Mofflumies to Chemin des Dames. Consisting of outposts, trenches, lines picquets, reinforced concrete bunkers, wide anti-tank ditches, and underground hospitals, fire control centres, offices, headquarters, all linked by telephones, with electric lights and water, provided with barbed wire and machine gun defences.

Hindernis. (G. *obstacle*).

Hinged drawbridge. A drawbridge introduced during the 14th century CE. The far end of the drawbridge was attached by ropes or chains to long beams of wood which were fitted with counter balances on their inner ends.

See 'bascule bridge', *'drawbridge'*.

Hira ishi. (J.). A type of broad and flat stone used to construct the main walls (*'ishigaki'*) of a castle but not the corners. Also known as a *'tsuki ishi'*.

See *'ishigaki'*.

Hirajiro. (J. *flatland castle*). A Japanese castle which was developed after the necessity of the *'yamajiro'* fortification declined with the cessation of civil war in the 17th century CE. These castles were built on the plains and served as administration centres of the surrounding area.

See *'hirayamajiro'*, *'yamajiro'*.
See Appendix 2. Shiro - a Japanese Castle?

Hirayagura. (J.). A simple turret, not tall enough to be a watch tower, normally used on the perimeter of an outer bailey (*'sannomaru'*), or used to connect 'yagura' to gates, larger *'yagura'* or *'tenshu'*.

See 'sumiyagura', 'yagura'.

Hirayamajiro. (J. *flatland hill castle*). A Japanese castle which was constructed on a hill on a plain, thus taking the positional advantage of the surrounding area.

See *'hirajiro'*, *'yamajiro'*.
See Appendix 2. Shiro - a Japanese Castle?

Hisn. (Ar.). Fortress.

Hitler Line. One of the German defensive lines built by Organisat Todt in Italy to slow the Allied advance (1944 CE), connected with the northern wing of the Gustov line to Terracu, provided with tank traps and concrete shelters.

See *'Gothic', 'Gustav' and 'Reinhard' lines.*

Hoard, Hoarding, Hourd. (Old French *hourt,* barrier). A covered wooden gallery built out from the parapet of a tower or curtain wall supported on corbels, providing for vertical defence of the area below thus reducing the amount of dead ground. The hoardings had a major disadvantage since they were made of timber, they were prone to firing, so they were eventually replaced by stone machicolations. In times of peace the hoardings could be removed.

See 'brattice (2)'.

Hochbunker. (G. *high bunker*). **See *'flaktower'.***

Höckerhindernisse. (G. hump obstacles). German WWII anti-tank obstacles consisting of re-enforced concrete in the shape of a pyramid, used in the 'Siegfried Line' and the '*Atlantic Wall*', in 1938 CE they were set in rows of four and in 1939 CE a fifth row was added.

See 'anti-tank obstacle', 'dragon's teeth', 'anti-tank pimple'.

Höckerlinie. (G. *cusp line*). The concrete anti-tank obstacles, also known as '*dragon's teeth*' were used to defend territory such as the German 'West wall'.

See 'dragon's teeth'.

Hochleitstand. (G. *high control station*). A German WWII multistoried concrete bunker used as an observation tower, fire control tower and a bunker. Extant towers can be found in Normandy (France) and the Channel Islands.

See 'leitstand', 'luftschutzraum', 'fire control tower', 'observation tower'.

Hochmotte. (G. *high motte*). **See *'motte'.***

Hochschloss. (G. *high castle*).

Hohenburg. (G. *hohen*, hill; *burg*, castle). A German castle built on a naturally strong site, usually a hill. All the defences were constructed on the lines of approach and the entrances. Also known as a '*German hill castle*'. Hohenburg came in two types; 1) '*Gipfelburg*' that stood on the summit of a hill with steep sides, and 2) '*Felsenburg*' that was built on a rocky outcrop. The chief advantage of a Hohenburg was its inaccessibility making it hard to assault but providing clear views for observation and fire of the surrounding area. Such sites had problems with supply of water and food, other supplies and exposure to poor weather.

See *'schildmauer'.*

Höhlenburg. (G. *cave castle*). A castle built into a cave, developed in the 11th to 12th centuries CE. Also known as a '*grottenburg*' (G. *grotto castle*). An example is Kropfenstein castle, Graubünden, Switzerland.

See 'corona', 'hohenburg'.

Hollandsche Water Linie. ***See 'dutch water line'.***

Hollow bastion. A bastion where the ramparts extend only along the flanks and the faces.

See *'bastion'.*

Hollow charge. (G.). Shaped demolition munition, used to penetrate at least 25cm (10 inches) of armour plate. Used by Germans to successfully attack the Eben Emael fortress defending the Albert Canal in 1940 CE, WWII.

Hollow traverse. A traverse fitted with passage.

See 'traverse'.

Holzpfählen. (G. *wooden posts*). **See 'obstacle', *'Rommel's asparagus'.***

Honjou. (J.).

(1) The main or central part of a castle, similar to the *'honmaru'* (compound).

See *'honmaru'.*

(2) The main or central castle of a group of castles, supported by a number of other minor castles. The *'honjou'* was often the residence of the lord of a domain.

See *'shijou'.*

Honmaru. (J. *main circle*). The main or inner fortified compound of a Japanese castle dating from the 16th century CE, which may consist of more than one compound, normally used as the residence of the lord. The *'honmaru'* could be flanked or surrounded by a second (*'ni no maru'*) and a third compound (*'san no maru'*).

See *'dôshinén', 'honjou', 'nawabari', 'ninomaru', 'renketsukei', 'san no maru', 'teikakukei'.*

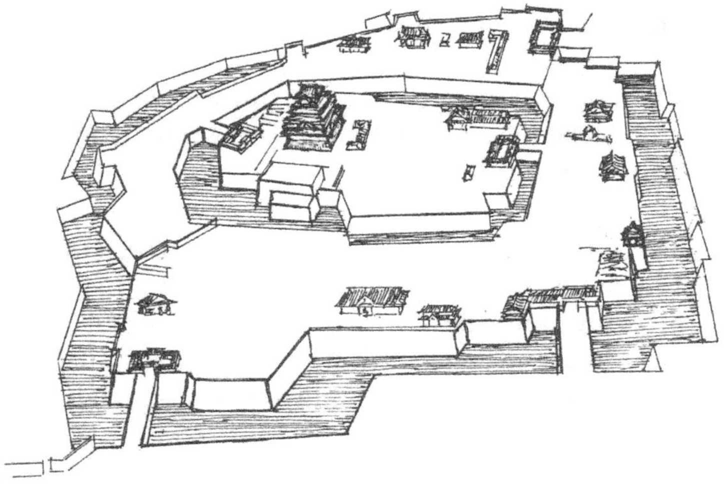

Honmaru and Ninmaru, Osaka castle, Japan.

Hooper loophole. A form of a loophole with its *'throat'* closed by a steel plate about 4 cm (1.5 inches) thick, pierced by a small opening for the rifle, used by the Japanese and Russians (WWII).

See 'gun loop', *'loop hole'.*

Hori. (J.). Trenches, ditches or moats of a castle.

Horikiri. (J.). A trench that cuts across a mountain ridge to impede attackers from easily climbing the ridge of a mountain top castle.

See 'bori', 'hakobori', '*kenukibori*', '*shojibori*', '*tatebori*', '*unebori*', '*unejotatebori*', '*yagenbori*'.

Horn.

See '*en bec*'.

Hornwork.

(1) An outwork consisting of two long curtains extending from the main works and was fronted by two '*demibastions*' connected by a short curtain, forming an outer bailey, it was smaller than a crown work but serving the same purpose.

(2) An earthen spur provided with banks and ditches.

Horse shoe. A small outwork, round or oval in shape, provided with a parapet.

Hot shot furnace. A specific oven used to heat cannon balls before firing them at enemy shipping.

Hourd, Hourding. See '*hoard*'.

Howitzer bunker. A bunker shaped so that the embrasure is wide and high enough for the full field of fire of a howitzer, developed in the 18th century CE.

See 'bunker'.

Huang or Gan Hao. (Ch. *dry moat*). **See '*hao*'.**

Hu cheng, (Ch. *protection wall*). A small auxiliary wall built at the base of the inner side of a main wall, taller than the 2 metres (6 ½ feet) of the Han cheng, designed to reinforce the main wall against damage from the frequent river floods.

See '*cheng*'.

Hu cheng zhe jian jia. (Ch. *wall protecting, arrow covering frame*). A special type of crenel shutter which hangs on mounts at a distance from the crenel, allowing defenders to lean out and attack an enemy at the base of the wall with some protection.

Huchette. (G.). A wooden shutter placed in a crenel which could be raised or lowered by turning an axle.

See 'crenel', 'crenel shutter'.

Húihúi pào (Mandarin, *trebuchet*). **See '*trebuchet*'.**

Hu men qiang. (Ch. *door protecting wall*). A solitary wall built in from the city gate, used to hide and protect the gate from artillery, usually built on top of a '*que tai*' some distance away from the city gate. Also known as '*dang men qiang*' (*door blocking wall*) or '*yong men*' (*block* *door*). During strife both sides of the '*hu men qiang*' could be blocked off with wooden palisades.

Hurdica. See *'hoard'*.

Hurdle. A basket used to hold earth, woven of willow, used in the building of defensive earthworks.

See *'gabion'*.

Hu xian qiang. (Ch. *danger protecting wall*). A fortified wall built near the bank of a moat with battlements, towers and gates but in a smaller size than the main city walls.

I.

Ichijõ-bekkaku. (J.). When two castles act as one in a supportive defensive role.

Ichinomon. (J.). The main gate of a castle. **See '*mon*'.**

Iejiro. (J.).

(1) A castle changing from a military role to that of a residential role.

(2) When a castle has a divided function, '*iejiro*' may only refer to the residential area.

See '*jiro*'.

Igel. (G. *hedgehog*). German WWII defensive system of mutually supportive strongpoints, and could be used for bases for counter attacks.

See 'box', 'island', 'stutzpunkt', 'widerstandnest'.

Ijang. A terraced hillfort of the Ivatan of the Batanes Islands of the Philippines, similar to the '*gusuku*' found in Okinawa, Japan, dating to the 12th century CE.

Impoundment. Flooded ground in front of or within a fortification for defensive purposes. Also known as '*inundation*'.

See '*ditch*'.

Improving. The continuing modification to a system of rapid entrenchment over time to make them safer, more comfortable and defensible, like widening the parapet, adding traverses, digging dug outs and fire pits behind the front line, adding supply caches and covered ways in the rear.

See '*entrenchment*', '*trench warfare*'.

Independent curtain walls. Each wall walk was only accessible by the doorways in each flanking tower which punctuated the length of the curtain wall, while these were weaker than the main gates, they were strong enough to withstand anything the attackers could bring up the scaling ladders. If an attacking force gained control of a wall walkway between two flanking towers there was little that could be done to improve their position if each curtain wall was separate from each other as well as from the adjacent towers.

During their effects to gain entry to the flanking towers or the inner defences they would be under fire from the defenders of the inner fortification as well as those of the flanking towers flanking the section of curtain wall they occupied, a most difficult position indeed.

See 'concentric castle'.

Indian merlons. Most Indian merlons are semicircular and point at the apex. Merlons were often provided with loopholes, and sometimes loopholes were provided under crenels.

See *'merlon'*.

Indrakosha. (Sanskrit). A plank of timber set up on a rampart between turrets with holes big enough to loose arrow through while providing protection to the archer.

Infanteriestuetzpunkte. (G. *infantry base*). A concrete structure built by Germany from 1900 to 1914 CE, which developed over time, with a ceiling thickness of 2 metres (6 feet 6 inches), covered with dirt, with rear access and a range of facilities for ammunition storage, observation, guard and dressing rooms, all connected by trenches.

Infantry shield. A small portable metal shield with a firing hole, used by infantry in WWI trench warfare.

See 'mantlet'.

In felsen gehauene. (G. *Carved into rock*). A rock enclosed position.

See 'sangar'.

Inner bailey. The bailey surrounding a keep; also known as the inner ward.

See *'bailey'*, *'ward'*.

Inner curtain wall. The inner curtain of a fortification surrounding the inner bailey.

See 'concentric castle', 'curtain'.

Inno koguchi. (J.). A type of defensive castle entrance which made it harder for an enemy to enter but also slowed down sorties of defenders by the use of barriers, such as '*umadashi*'.

See *'koguchi'*, *'umadashi'*, *'youno koguchi'*.

Inno nawa, Inno shiro. (J.). Defensive designed castles with defensive designed compounds, bridges and barriers sited at the rear of a castle.

Inskonce.

(1) To fortify, to entrench with sconces.

See *'ensconce'*.

(2) An earthwork consisting of a ditch and rampart stretching the length of a frontier, which was interspersed by bastioned and fortified villages.

Interior crest. The line formed by the meeting of the interior and superior slopes of a parapet or rampart.

Interior flanking angle. The angle formed by the junction of the line of defence and the curtain.

Interior side. The line from the centre of one bastion to the centre of its neighbouring bastion.
Interior slope. The earthen slope on the inside of a rampart or parapet.
Internal polygon. The line followed by the curtain which runs from the gorge to the base of a bastion, and is parallel to the external polygon.
Intervallum. The space between the wagons used in the form a '*wagenburg*' and the infantry and cavalry stationed in the centre of the field defence work.
Intra-mural timbers. Large timbers placed within stone walls to add strength.

See '*crib work*'.

Inubashiri. (J. *dog run*). A strip of land between the ramparts of a castle and the moat used as a pathway for patrols.

See '*chemin de ronde*'.

Inyou no kuruwa. (J.). An offensive castle compound,
Irregular works. See '*regular works*'.
Irish battlement. A battlement typified by the use of stepped merlons, where a smaller merlon rises out of a larger main merlon.

See '*stepped merlon*', '*venetian battlements*'.

Isama. (J.). A loophole or gun port set in a castle's wall set at shoulder height of a seated defender, most often in the form of a small circle but also in squares, upright rectangles and triangles.

See '*sama*'.

Ishigaki. (J.). A dry stone wall used in the construction of a Japanese castle. The '*ishigaki*' was introduced into Japanese military architecture because of the increase in the frequency of conflict, and the earthworks were no longer a form of effective defence. The technique of construction involved embedding stones into an inclined earthen embankment and locking them in place with smaller stones ('*meji*'). This allowed the whole wall to shake and thus dissipate the effect of earthquakes, which are a normal occurrence for the country. The '*ishigaki*' were used as a foundation for such structures as; '*tenshu*', '*yagura*' and '*dobei*'.
The most frequently used stone for '*ishigaki*' was andesite, granite, quartz, porphyry and gneiss.
There are three types of shapes of '*ishigaki*' depending on recline angle, 1. The walls are at 45 degree elevation with the top stones are vertical ('*narashi*'), 2. The walls recline at 50 degree elevation and the top 20% of stones are vertical, the third type has an incline of about 80 degrees,

the wall slopes upwards in a concave curve, with the top 25% completely vertical.

There are two types of *'ishigaki'* of construction; the first usually involves the pounding and insertion into the soil of the ishigaki (*'uchikomihagi')*, while the second type involves the cutting and insertion of masonry (*'kirikomihagi)*.

There are a number of ways of building an 'ishigaki', depending on the shape and placement of stones.

See *'nobori ishigaki'*, 'kikkozumi', *'kirikomihagi'*, *'nozurazumi'*, *'nunozumi'*, *'ogi no kôbai'*, *'ranzumi'* 'tamaishizumi',*'tanizumi'*,*'uchikomihagi'*.

Ishi-otoshi, Ishi-otoshi mado. (J. *stone drop or stone dropping window*). A structure which flares out from the wall, which was used for the vertical defence of the base of the walls of a Japanese castle, consisting of a section of the wooden or plastered wall of a; *tenshu*, tower, connecting tower or other structures, which was built out over the stone walls at strategic locations. The structures enabled the defenders to shoot guns, loose arrows as well as throw stones at attackers attempting to scale the walls but was too thin to enable an attacker entry. The aperture could be closed for protection if necessary, by a plank of thick wood which could be strengthened by iron bars.

See '*box machicolation*', '*machicolation*'.

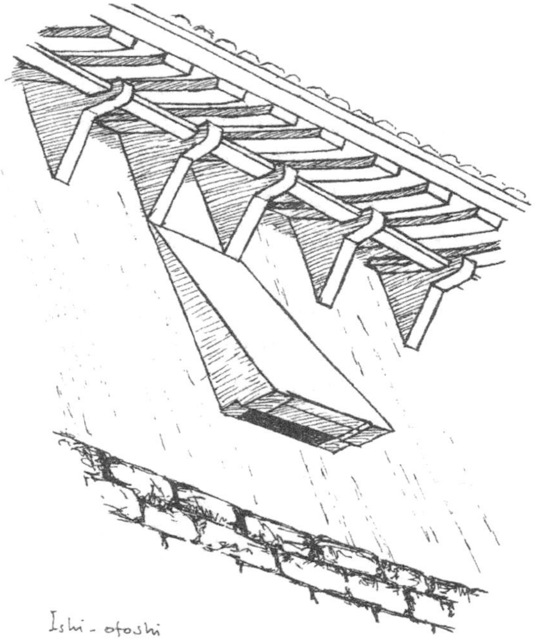

Ishisama. (J.).
(1) A loophole cut into the top of a stone wall, under the fence on top.
(2) Loopholes made through the stone walls of a castle.
See 'hako hazama', 'sama', 'shinogi hazama', *'taihosama', 'tepposama', 'yasama', 'zama'.*

Ishiuchidana. (J.). A platform which runs along the interior wall of a castle ('*hei*') on which piles of stones are provided for throwing at the enemy.

Ishi uchi tana. (J. '*stone throwing shelf*'). The planks of wood placed on timbers jutting out of the inner face of the walls of Japanese castle for archers and gunners to stand on to loose arrows and fire guns at an enemy.

Ishižumi. (J.). Stone-piled walls.

Island. British WWII enclosed fortified positions developed as part of invasion preparation in 1940 CE, provided with emplacements, dugouts, connecting trenches and tunnels, along with road blocks, barricades and obstacles.

See 'igel', 'stop lines'.

Italian machicolation. A machicolation which was corbelled out on a long brick projection descending down a third of the tower height, forming ribs around the upper sections of towers and curtains, and joined at the top by semicircular arches.

See *'machicolation'*.

Italian merlon. A merlon provided with a curved indentation at its apex. Also known as a *'notched merlon'*.

See *'merlon'*.

Italian outline, Italian trace.

See *'bastion fort'*. *'bastion system'*, *'trace'*, *'trace italienne'*.

Iujiao. See *'teng pai'*.

Iya, Iyanuwo. See *'Benin moat'*.

J.

Jala-durga. (Sanscrit. *water fort*). Also known as '*audaka durga*' or '*ab durga*'.

See 'durga'.

Jambs. (Fr. *jambe*, leg). The wings attached to a tower house of the English/Scottish border region, which were used to increase the flanking ability. The design developed into what is known as a '*Z-plan towerhouse*'.

See 'towerhouse', 'Z-plan towerhouse'.

Jayb. (Ar.). An underground cistern in a fortress.

Jiao tai. (Ch. *corner platform*). A tower or bastion built at the junction of two walls. Also known as a '*Wan jun tai*' (*army lookout platform*). A '*Kong jun tai*' is a stand-alone tower built close to a '*jiao tai*', a watch tower.

Jiao lou. (Ch. *corner tower*). A heavily fortified corner bastion, similar to the main gate tower of a city because of the lack of mutual support on either side of the wall.

See 'di lou'.

Jian chuan. (Ch. *arrow window*). A large loophole of an enclosed or roofed structure such a '*di lou*' or '*di tai*'. The '*jian chuan*' of a '*dia lou*' had window hatches, whereas on a '*di tai*' they did not.

Jian yan. (Ch. *arrow hole*). **See '*yan*'.**

Jinya. (J.). Local administration centre for a small domain, which could be fortified with moats, embankments and watch towers, some of which were left over from an earlier castle on the site.

Jô. (J.). The proper noun suffix in Japanese for the word '*castle*'. The term comes from the Chinese character for city and wall ('*shiro*' or '*jô*'). Later, the meaning changed from city to castle; for in Japanese a '*walled enclosure*' means a castle rather than a city.

See* 'shiro'*, Appendix 2. Shiro - a Japanese Castle?

Jôkamachi. (J. *town below the castle*). The town or city which developed about a Japanese castle, which was carefully set out in designated zones according to rank; the samurai residences were situated nearest to the castle, and then came those of the merchants and then those of the artisans. Thus, the '*jôkamachi*' became centres of trade and growth, centred on the defensive works of the castle.

Joukamachi. (J. *town below a castle*). A town located at the base of a mountain castle, and could develop into a castle city ('*jouka no takai*'). Also known as a '*negoya*'.

Juliet. A type of circular donjon. **See 'donjon'**.
Juma. **See '*teng pai*'**.

K.

Kabal. USA and Coalition bases consisted of earthworks in the northern part of Kuwait, containing dining and recreation facilities, air conditioned tents, storage for arms, tanks and armoured vehicles.

Kabukimon. (J.). A type of roofless gate consisting of two gate posts connected by a horizontal beam dating from the Kamakura period, the gates swung inwards, a small side door may be provided called *'kowakitobira'*.

See '*mon*'.

Kaiserburg. (G. *kaiser*, emperor; *burg*, castle). The German imperial castle of the 12th and 13th centuries CE.

Kakehashi. (J.). A removable wooden bridge supported on posts over a moat, handy in times of siege to deny enemy access.

Kakushiguchi. (J.).

(1) A hidden or inconspicuous entrance to a castle through an earthen embankment beneath a castle wall.

(2) A bent castle entrance between parallel ramparts and a moat, forming a small compound, thus the entrance is obscured by the external observer.

See '*bent entrance*'. '*koguchi*', '*masugata mon*'.

Kakushiguruwa. (J.). A hidden compound or bailey of a castle.

See '*kuruwa*'.

Kakushisama. (J.). Loopholes of a castle which were not discernible from the outside.

See '*sama*'.

Kakushizama. (J.). Loop holes closed by a door or plug that matched the surface on the outside, making it hard to detect.

See '*sama*'.

Kaku umadashi. (J.).

(1) A rounded shaped earthen embankment or wall built in front of a Japanese castle gate.

(2) A square shaped wall or earthen embankment built in front of the main castle gate, dated from the 17th to the 19th century CE.

See '*barbican*', '*doi*', '*hori*', '*koguchi*', '*maru umadashi*', '*umadashi*'.

Kamae. (J.). An earthen embankment or earthworks, and has the same meaning as *'kuruwa'*.

See '*doi*', 'kuruwa'.

Kammburg. (G. *ridge castle*). An uncommon German castle built on the crest of a ridge, which could be attacked from both sides of the ridge.

See *'castle'*.

Kammertor. (G. *chamber gate*). A type of German gateway used on medieval town or castle gates, consisting of two successive gates, connected by a defensive passageway, which can connect two enceintes, provided with *'porticullis'* and *'drawbridges'*. Chamber gates can be found in *'zwingers'*.

See *'gatehouse'*, *'zwinger'*.

Kampfstand. (G. *fighting stance*). A German bunker which could be armed with a range of weapons from; anti-tank guns, flak guns, machine guns, mortars, or field artillery or from naval vessels.

Kandvarini. (Sanskrit). A thin wall on the parapet for the protection of the garrison.

Kane no umadashi. (J.). A right angled barrier at a castle entrance located at an indented corner of a castle.

See *'koguchi'*, *'umadashi'*.

Kannuki. (J.). A gate bar used to secure a gate, usually a heavy timber covered with iron plates nailed on, which was slid through cleats (*'kannukikasugai'* or *'kannukimoch kanamono'*) to lock or unlock the gate.

Kano yagura. (J.). Corner turret.

See *'yagura'*.

Kapata. (Sanskrit). The door or doors of a gate of an Indian fort, which was often protected by iron spikes (See *'parighas'*) to prevent attack by elephants.

Kapishirsha. (Sanskrit). A turret constructed at the top of a parapet, resembling the head of a monkey. Also known as a *'kangoora'*.

Karabori (J.). Trench or dry or empty moat, usually with *'V'*-shaped cross section, most often seen in mountain castles (*'yamajiro'*).

See *'bori'*, *'mizubori'*, *'yagenbori'*.

Karamete. (J.). The rear of a castle.

See *'oote'*.

Karametemon. (J.). The rear gate of a castle.

See *'mon'*, *'ootemon'*.

Kasane umadashi. (J.). Multiple naturally occurring or constructed barriers protecting a castle's entrance, usually placed one in front of another, including mountain ridges and cliffs. Usually employed when there is no interior gate complex.

See *'umadashi'*.

Kasanezake. (J.). Parallel paths used to ascend or descend from an earthen embankment (*'doi'*) or stone wall (*'ishigaki'*). Used to change troops at the same time, thus making the movement more effective.

See *'mushabashiri'*.

Kasbah. A fortified village, a traditional pre-Saharan habitat made of clay in Morocco (i.e. Ait Benhaddou).

Kastro. (Gr. *Κάστρο*). Castle.

Kastron. A late Byzantine fortified town.

Kata yagenbori. (J.). A moat with steep sides, where the bottom of the moat is not in the centre. This increases the angle of the interior slope of the moat (*'yangen'*, a mortar used to grind herbs).

See *'bori'*, *'yangebori'*.

Kawara-dobei (J.). A type of earthen wall (*'dobei'*) of a castle, with tiles embedded in the wall.

See *'dobei'*.

Kazashi. (J.). Items such walls, embankments, even plants sited outside a castle to screen offensive operations.

See *'shitomi'*.

Kazashi no dote. (J.). An earthen embankment raised inside the gates of a castle to block the enemy's view into the castle.

See *'doi'*, *'dote'*, *'nagadote'*.

Keel. The pointed base of a fortification's walls.

See *'battered base'*, *'plinth'*, *'spur'*, *'talus'*.

Keep. The English term for a donjon; the inner tower of a castle, usually the strongest, used by the besieged as the last refuge tower. By the 12th century CE a keep was used as more of a residence for the lord or castellan.

See *'hall keep'*, *'shell keep'*, *'keep house'*, *'tower keep'*.

Keep and bailey castle. The castle type was developed from the *'motte & bailey castle'* of the Normans, dating from the late 12th and early 13th centuries CE. Consisting of a stone keep surrounded by a walled enclosure, the keep either incorporated or replaced the motte. Two types of keep were used and they were; the *'shell keep'* and the *'square keep'*. This kind of castle however did not present a co-ordinated defensive system and thus had a noticeable military weakness.

Once the bailey was taken by an enemy, the next part of the operation was to take the keep. The operation was made easier by the fact that the two components did not support each other; once the bailey fell to the enemy, all the enemy's forces could then be concentrated on the keep.

See *'bailey'*, *'keep'*, *'motte'*, *'shell keep'*, *'square keep'*.

Keep gatehouse. A fore work designed to act as the defensive kernel of a concentric fortress, as opposed to a keep. It was strong enough to hold out independently when all else had fallen. Typical of the Edwardian concentric castles.

See *'great gatehouse'*.

Keep house. A keep containing all the living quarters as well as being a serviceable defensive fortification.

See *'keep'*.

Keepless castle. A castle introduced by the Crusaders, which consisted of a strong curtain wall provided with powerful flanking towers, a well defended entrance, and no actual keep.

See 'courtyard castle', 'quadrangular castle', 'quadrilateral castle'.

Kelheim-style. ***See 'pfostenschlitzmauer.***

Kenukibori. (J.). Shallow '*V*' shaped trenches used to defend Japanese castles.

See 'bori', 'hakobori', *'horikiri'*, *'shojibori'*, *'tatebori'*, *'unebori'*, *'unejotatebori'*, *'yagenbori'*.

Kernburg. The inner ward of a castle or fort.

See *'burg'*, *'ward'*.

Kernmotte. (D. *kern*, core). A transitional motte, where a motte was added to an already existing ringwork, the ringwork became the bailey.

See *'motte'*.

Kernwerk. (G. *core work*).

(1) A small fort within a German 19th century CE fort, like a castle.

See 'réduit'' 'redoubt'.

(2) A cluster of defence works in the middle of a '*festung*', consisting of bunkers with machine guns, artillery, barracks, stores, generators and hospitals and the like.

See 'festung'.

Kettle emplacement. A shallow dish shaped gun position.

See 'bettung', 'gun pit'.

Keyhole gun port. A gun port shaped like an inverted keyhole, with a round hole for the gun at its base, and above it was a slit used for sighting and as a vent for the gases produced from firing the cannon.

See *'gun loop'*, *'gun port'*, *'loop'*, *'port'*.

Khanjana-durga. (Sanskrit). An Indian fort sited on a marsh that is surrounded by a thorny forest.

See 'durga'.

Ki. (J. *rice plant fort*). A fortification used as a defence against an enemy's archers, consisting of a defensive palisade made of bundles of rice plants piled together, dating from the 6th century CE. This type of fortification was used by the Mononobe clan (who were against the introduction of Buddhism) against the Soga clan (who were advocates of Buddhism).

Kido. (J.). A gate leading to a castle's compound or an enclosure containing samurai housing, in the 13th to 16th c CE. The *'kido'* consisted of two gate posts sunk into the ground (*'hottatebashira'*, post hole) which support two timber swinging doors (*'hirakitobira'*, swing door).

Kikkozumi. (J.). A type of *'kirikomihagi'* ishigaki where the stones are precisely cut and fitted together with five or six sides.

See *'ishigaki'*, *'kirikomihagi'*, 'nozurazumi', *'nunozumi'*, *'ranzumi'*,*'uchikomihagi'*.

Killa. (Marathi). Indian fort.

Killing ground. An area surrounding or part of a fortification, which was deliberately kept clear to deprive an enemy of any cover, thus exposing the enemy to a greater amount of fire, with little chance of survival.

See *'glacis'*, *'independent curtain walls'*.

Kirigishi. (J. *bluffs*). Bluffs used as basic defence for an early mountain castle.

Kirikomihagi. (J.). Ishigaki where the stones are precisely worked and there are no gaps between the stones, there are drainage holes provided. Used for the steepest castle wall with an incline of 80 degrees in a concave curve, with the top twenty five percent vertical, used in the early modern period for front entrances.

See *'ishigaki'*, 'nozurazumi', *'nunozumi'*, *'ranzumi'*, *'uchikomihagi'*.

Kleinschartenstand. (G. *small embrasure stand*). A small and inexpensive well armed casement built during the late phase of construction of the *'Atlantic Wall'*.

See 'Atlantic Wall', 'casemate'.

Klinkett. A small gate in the palisade enclosing a fortification.

See *'guichet'*, *'postern'*, *'wicket'*.

Knife edge. A portable barrier consisting of barbed wire coiled round a wooden pole supported on either end by a cross of timber.

See *'barbed wire'*.

Kochbunker. (G.). A German WWII prefabricated circular concrete observational or combat pit for a single soldier, often part of a trench system. Also known as *'K-Rohrenstand'*.

Kõgo-ishi. (J. *stones of divine protection*). An early Japanese fortification, dating from the 4th to the 6th century CE, the Asuka period, particularly around Fukuoka, consisting of rows of stones encircling a hilltop periodically interspersed with gateways and sluices, these defences reinforced the natural defences of the site.

Koguchi. (J. *small entrance*). Gate posts or a bent castle entrance, defended by '*sama*' and '*yagura*'. A narrow winding entrance with unaligned walls was known as a '*kuichigai koguchi*'. A later form developed with an open courtyard walled on the two sides, with the main gate on the adjacent side leading the inner part of the castle, called a '*masugata koguchi*'.

See ***'kôrai mon', 'masugata mon', 'yojoya'.***

Kojiro. (J. *little castle*). A subsidiary castle to the main castle '*oyajiro*' (*parent castle*).

Kong jun tai. (Ch. *army commanding platform*). A stand alone tower built close to a '*Jiao tai*' (Ch. *corner platform*) connected to a curtain wall by a bridge. Also known as a '*wang jun tai*' (Ch. *Army lookout platform*). An example can be seen on the northwest corner bastion of the rebuilt city wall of Datong, China.

See 'albarrana', 'jiao tai'.

Kong xin di tai. ***See 'di tai'.***

Kôrai mon. (J.). The smaller initial gate of a '*masugata mon*' or barbican gatehouse complex of a Japanese castle, consisting of a three roofed structure (a central roof was flanked by another two roofs). This gate restricted access to the courtyard beyond, and if the enemy did gain entry to the courtyard they would be confined within that space and would be exposed to the fire of the defenders. The older style of '*korai mon*' had a roof with equal height surrounding walls, the newer style of gate was taller so the roof is higher than the surrounding walls.

See '*masugata mon*'.

Masugata mon and Korai mon.

Koshikuruwa. (J. *waist compound*).

(1) A terraced bailey or sub-bailey found on the side of a mountain castle (*'yamajiro'*).

(2) On flatland castles (*'hirajiro'*), the long narrow compounds constructed lower than the main compound were also known as *'koshikuruwa'*.

See *'kuruwa', 'sodekuruwa'*.

Kota. Muslim Filipino fortresses built in stone and bamboo, surrounded by trenches. Used to protect family groups and lords alike, used to block the Spanish advance (16th and 17th century CE) into the region. Also known as *'moong'*.

Kotenshu. (J.). A small keep or subsidiary tower.

See *'tenshu'*.

Kotta mara. A fortified raft or battery from Borneo, dated from the Banjarmasin War (1859 – 1906 CE) used to defend waterways.

Kouraimon. (J. *Korean gate*). Served as a small outer castle gate, usually at the front gate, leading to a *'masugata'*, then a *'yaguramon'* making a *'masugatamon'*. Also known as a *'kabukimon'*.

See *'kôrai mon'*.

Kremlin. The citadel of a Russian city made of stone, earth and timber, such as Novgorod or Moscow.

Ksar. (Ar.). A type of fortified Northern African village or fort, normally associated with the Berbers.

See *'qasr'*.

Kuichigai. (J.).

(1) Cross-over embankments.

(2) Where one of the embankments (*'doi'*) on either side of a castle entrance (*'koguchi'*) juts out in front of the entrance.

See *'doi'*, *'koguchi'*.

Kugurido. (J. *subterranean*). A small gate set in a larger gate, provided with a high threshold and low lintel, allowing only one person at a time to pass, acting to restrict the number of people using the gate, thus increasing the security.

See *'mon'*, *'wicket'*.

Kuklos. (Gr.). A field fortification made of kneeling baggage animals (i.e. camels).

Kumbhalgarh. ***See 'Great wall of India'.***

Kuri ishi. (J.). Small stones used as packing or backfill for the larger stones of the *'ishigaki'*.

See *'ishigaki'*.

Kurogamae. (J.). A black earthen embankment built high enough to not allow observation of the buildings inside a castle.

See *'doi'*, *'kamae'*, 'kuruwa'.

Kurumabashi. (J.). A drawbridge leading to the inside of a citadel. **See *'hikuhashi'*.**

Kurawa. (J.). A general term meaning a Japanese castle's compound or enclosure, surrounded by a moat, earthen or stone wall. Also known as a *'maru'*.

See*'honmaru'*, *'ninomaru'*, *'sannomaru'*.

Kuruwa. (J.). A bailey or group of baileys of a castle. When making a *'yamajiro'* (mountain castle) the area made flat and bounded by earthworks was called the *'kuruwa'*, similarly for a *'hirajiro'* the area bounded by the spoil from digging the moat also was called the *'kuruwa'*. Kuruwa came in varied types depending on shape and placement (i.e. *'deguruwa'* = projecting bailey, *'kigash-kuruwa'* = east bailey, *'nishi-kuruwa'* = west bailey, and so on).

See*'dekuruwa'*, *'koshikuruwa'*, *'obikuruwa'*, *'shukuruwa'*.

Kugelstand (Kugelbunker). (G.). Special structure made from reinforced concrete with a hole at the top for a German (WWII) machine gun nest, used in the defence of the Slovakian frontier.

Kumbhalgarh. See *'Great Wall of India'*.

Kuruwakubari. (J.). The distribution of compounds of a Japanese castle is known as *'kuruwakubari'*.

See 'kuruwa'.

Kushk. A fortified rural villa of the *'Sudg'* (Southern Uzbekistan and north-western Tajikistan).

Küstenvorfeldsperren. (G. *coastal apron barriers*). A German beach obstacle. Also known as '*K-Sperre*' or '*Vorstrandhindernissen*'.

See 'beach obstacles'.

Küver. (G.). A bunker of the Luftwaffe in Holland that were part of the coastal defences of the Atlantikwall in World War II.

See '*Atlantic wall*'.

Kyokan. (J.). The residential area of a castle.

Kyuukado sama. (J.). A gunport or loophole in the corner of a castle wall providing a larger range of firing than in a straight section of wall.

See '*sama*'.

L.

Laager. A circular encampment of wagons used by the Boers of South Africa against Zulu attacks, 1830's CE.

See '*wagonberg*'.

Laboratory. A place in 19th century CE fortification where gunpowder was made and placed into shells and grenades, before being moved to the magazine.

Landesberg, Landesburgen, Landesherrliche berg. (G. *state castle*). A large German castle dating between the 8th and 10th century CE, located east of the Rhine river, used for control of the state.

Land front. The front of a fortification designed to defend against land based attack.

Land walls. Linear defence works delineating a border consisting of ditches, ramparts, walls, towers and gates. Examples are; Antonine (Scotland), Atlantic (Various) Constantinople (Turkey), Danewerk (Denmark), Great Wall of China, Hadrian's (England), Maginot (France), Offa's dyke (Wales).

See 'linear fortifications'.

Land walls of Constantinople. See '*Theodosian Land walls*'.

Landwehr. (G. *land defence*). A German term for border defences or demarcation and enclosures in Central Europe, using earthworks, dykes, ditch (G. landgraben, land ditch) and hedging (G. landhege, land enclosure) used for protecting towns and villages in the mid to late Middle ages, which could run over a hundred kilometres long.

See '*Danewerk*', '*limes*'.

Lascaris Towers. (Maltese, *Torrijiet ta' Lascaris*). A series of mostly small coastal watchtowers built in Malta by the Order of Saint John between 1637 and 1652 CE. Mostly square in plan, inward sloping sides at the base to straight sides to the top.

Latera praetorii. (L. *the sides of the hall*). The central area of a Roman fort, between the *via principalis* and *via quintana*, where the headquarters (*principia*), the commanding officers house (*praetorium*), the granary (*horreum*) and sometimes the hospital (*valetudinarium*) were located.

Lavra. Byzantine monastic retreat, sometimes fortified.

Laws of War. Developed over time to cover the conventions and agreements between combatants, which included siege warfare, this included; parley, truce, the terms of surrender, hostages, safe conduct, the prospect of relief and what would happen when that did not happen in a determined time frame, the treatment of the wounded and dead. Later these laws were developed into the '*Geneva convention*'.

Leitturm. (G. *lead tower*). A German tower (WWII), used for fire control, command, listening bunker or as a small flak tower. Also known as a '*L-tower*'.

See *'fire control tower', 'flak tower'.*

Legionary fort. A Roman fort used to house a legion, which was planned out with defences (lilli, ditches, ramparts, palisades, towers), gates, roads, barracks, baths and toilets, granary, stores, workshops, hospital, headquarters etc, built in stone. Examples in Britain include; York, Caerleon and Chester.

See *'Agger', 'Ascensi valli', 'Antemuralia,' 'Ballistarium, Burgi', 'Cardo maximus', 'Castella', 'Castellum', 'Castra', 'Cervi', 'Cespites', 'Cippi', 'Decumanus maximus', 'Foss' 'Fossa fastigata', 'Fossa punica', 'Latera praetorii', 'Lilia', 'Lorica', 'Numerus', 'Porta decumana', 'Porta praetoria', 'Porta principalis', , 'Porta principalis dextra', 'Porta principalis sinistral', 'Punic ditch', 'Spina', 'Vallum', 'Via decumana', 'Via quintana', 'Via praetorian', 'Via principalis', 'Via sagularis', 'Via vicnariae'.*

Leitstand. (G. *control centre*). A fire control centre.

Lessor or Minor fieldworks. Earthworks constructed within an interior or rear ditch for the soldiers to fire over the parapet while standing in the ditch.

See *'ditch', 'fieldwork'.*

Licence to crenellate. A warrant of royal permission to build a castle or to fortify an existing structure such as a manor house (an example is from 1482 CE, the '*Licence to crenellate*' in which Edward IV gave Sir Edmund Bedingfield permission to build a battlemented house at Oxburgh Hall.

See *'adulterine castles'.*

Light anti-aircraft emplacement. An Australian circular emplacement from WWII, consisting of a wall of earth-filled 44 gallon (166 litre) drums, augmented with a layer of sand bags. A Lewis gun was mounted on a metal swivel on top of a wooden upright stuck into the ground in the centre of the emplacement. An example can be found in Batchelor, Northern Territory, Australia.

Ligne Maginot. ***See 'Maginot Line'.***

Ligne Morice. A French barrier along the Algerian- Tunisian border, dating to 1957 CE, consisting of an electrified fence, barbed wire entanglements and minefields, supported by searchlights, radar sensor and artillery batteries.

Lilia. Circular pits provided with sharpened stakes at the bottom in addition to the defences of a Roman fort. Examples were found at the Antonine wall fort at Rough Castle, Scotland, UK.

See 'trap holes', *'trous-de-loup'*, *'wolfsgrube'*.

Limes, limites. The frontier barriers of the Roman Empire, using natural barriers where possible, and a series of *'castra stativa'* linked by ramparts topped by palisades (i.e. Germanic, Hadrian's Wall, Antonine Wall).

See 'Antonine wall', 'foss', 'burgi', 'Fossatum Africae', 'Hadrian's wall', ' Limes Germanicus', 'limes Raetian', 'specula', 'vigilarium'.

Limes Arabicus. *See 'strata diocletiana'.*

Limes Germanicus. The fortified northern border of the Roman Empire on mainland Europe, at its full extent went from the North Sea outlet of the Rhine to near Regensburg (Castra Regina) on the Danube, which was provided with ditches, rampart, towers and gates. Over a period of time it was moved and repaired. In 83-85 CE, Saalsburg to Butzback 50km (30 miles) – series of watch towers connected by a track, later a light fence was added. Danube River to the Black sea (Trajan's wall), 50 km (30 miles). Later extended to reach Lahn and Main rivers 120km (75 miles). In 117 CE, realignment and reconstruction by Hadrian's command for an eastern extension from Mittenberg south to the Neckar and Danube rivers, 280km (175 miles) with fifty five Auxiliary forts and thirty forts in reserve.

See 'Antonine wall', 'Hadrian's wall', 'limes', 'limes Raetian'.

Limes Raetian. The fortification on the frontier of the Roman province of Raetia, extending from Lorch to the Danube River about 550 kilometres long (342 miles), in its latest form consisted of a stone wall with built in watch towers, without a palisade, and forts behind.

See 'Antonine wall', 'Hadrian's wall', 'limes', 'limes Germanicus'.

Lincolnshire three-bay pillbox. A British WWII pillbox located only in Lincolnshire, rectangular in plan, consisting of a square roofed area with an embrasure in each of the outward faces at each end, and a central raised platform for anti-aircraft defence. There are 44 examples extant, including an example in Saltfleetby, Lincolnshire, England.

See 'British hardened field defences', 'FW3', 'pillbox'.

Line. A work raised by an army extending from one position or post to another, the army encamped behind the work and used it to guard the territory the work traversed.

Line capital. A line drawn on a map of a fort from the angle of the gorge to the angle of the bastion.

Line cogrital. A line drawn on a map of a fort from the angle of the centre to the angle of the bastion.

Linear fortifications. Long fortifications which could take various forms varying from simple ditch and rampart to heavily fortified frontiers. Basically linear fortifications were static land based defences that needed to be defended, maintained and where necessary upgraded, they were expensive to build and maintain, their longevity can be overcome by changes in circumstances (the fall of the Roman empire, *'Hadrian's Wall', 410 CE*), changes in politics or borders (*'Danwerk'*, 650 CE to the 19th century CE), changes in types of warfare (German *'blitzkrieg'* of Poland, 1939 CE) or by passed (Germans attacking the French via Belgium bypassing the expensive and well fortified *'Maginot line'*. 1940 CE), they could be overcome by an overwhelming force (the Allies' Normandy landings, *'Atlantic Wall'*, 1945 CE) or by changes in fire power (the cannons of the Turks, *'Theodosian Land walls'*, 1453 CE).

See 'Alpine Wall', 'Anastasian wall', 'Antonine wall', 'Atlantic Wall', 'Baltic defence line', 'Bar-lev-line', 'Barrière de fer', 'Benes Lines', 'Berlin wall', 'Cherta', 'Çakmak Line', 'Circumvallate', 'Constantian Wall', 'Contravallation', 'Corradino Lines', 'Cottonera Lines', 'Crimean defence line', 'Danevirke', 'dike', 'fence (fortified)' 'Floriana Lines', 'Fossatum Africae', 'Götavirke', 'Gothic lines', 'Great wall of China', 'Great wall of Gorgan', 'Great Wall of India', 'Great wall of Korea', 'Great wall of Qi', 'Great wall of Sindh', 'Gustav line', 'Hadrian's wall',

'Herclian Wall', 'Hexamilion wall', Hilsea Lines', 'Hindenberg Line', 'Hitler Line', 'Land walls', limes', 'limes Germanicas', 'limes Raetian', 'Lines of communication', 'Lines of cremaillere', 'Lines of Torres Vedras', 'Maginot Line', 'Mannehiem Line', 'Mareth Line', 'Metaxas Line', 'Moesian Limes', 'Molotov Line', 'Morice Line', 'National Redoubt', 'Offa's dyke', 'Outer London defence ring', 'Panther-Wotan Line', 'Perekop wall', 'Pré carré', 'Promenthouse fortified line', 'Reinhart lines', 'Rupnik Line', 'Russia-Urkraine wall', 'Santa Margherita Lines', 'Siege lines', 'Siegfried line', 'Silesian walls', 'Skäne line', 'Stalin line', 'stop lines', 'strata diocletiana', 'Surovikin line', 'Switch lines', 'Südwall', 'Theodosian Land walls', 'Victoria lines', 'Wagner line', Waterline', 'Wat's dyke', 'Zasechnaya cherta'.

See Appendix 4. Linear fortifications.

Line of approach. A trench dug at right angles from the covert way by the besieged garrison so as to enfilade an enemy's position.

Line of circumvallation. Earthworks that consist of a parapet with a ditch in front, punctuated by redans which flank the parapet, built in conjunction with lines of contravallation, and used to defend a besiegers camp and positions against a relieving army.

Lines at intervals. A line of mutually supportive detached works, such as *'lunettes', 'redans', 'redoubts'*, that were not connected by parapets. Also known as a *'discontinuous line'*.

See *'detached works'*.

Lines of bastions. A succession of lines of defence consisting of bastioned curtain walls (e.g. those of a concentric castle).

Lines of communication. Defensive lines which were built around the City of London (England) between 1642 and 1643 CE by the Parliamentarians to protect London from the attacks of the Royalist armies of King Charles the 1st. The lines consisted of ditches and earthen ramparts connecting twenty three fortifications of various types including *'star forts'*.

Lines of contravallation. See *'contrevallation'*, *'counter vallation'*.

Lines of cremaillere. Lines of defence consisting of alternating short and long faces at right angles to one another.

See *'cremaillere'*.

Lines of defence.

(1) Any artificial, semi-natural or natural obstacle which can be used to impede an enemy's approach (e.g. fortification, ditch, terrain, etc.).

(2) The distance between a bastion's salient angle and the opposing flank; the bastion flank.

(3) A series of fortified outposts.

Lines of tenailles. A series of salient and re-entering angles formed by parapets.

See '*tenailles*'.

Lines of Torres Vedras. Portuguese defence lines consisting of three lines of mutually supportive batteries and redoubts linked by trenches, defending the harbour of Lisbon against the French, 48 km (30 miles) long, dating from 1810- 1811 CE.

Lis, Liss.

(1) A form of defence similar to a palisade consisting of upright stakes tied together by vines and twigs.

(2) An Irish fortification consisting of a circular mound surrounded by a ditch, used as protection for a village.

Lisiére, Liziere. The berm or walkway along a parapet.

See '*foreland*'.

Listening posts.

(1) Located in a sap trench about 30 metres (about 98 feet) forward of the '*front line trench*'. Used to monitor enemy activity including mining.

(2) Concrete (Acoustic Mirrors) dish structure used to detect incoming aircraft, (i.e. England, early part of WWII, superseded by Radar, eastern side of Dungeness).

Listening well. A shaft dug down into the ground during a siege to detect an enemy's attempt to tunnel under an entrenched position.

See '*mine*'.

Lists. (Teut.). An area set aside for jousting, which was incorrectly applied in the 19th century CE to the spaces between two lines of curtain wall or a bailey in the mistaken belief that jousting took place there.

See ‘*ward*’.

Litus Saxonicium. *See ‘Saxon shore forts’.*

Lodgement. A work made by besiegers in the part of a fortification taken by their actions, and used to hold it against counter attack.

See ‘*beach head*’.

Logistic bunkers. Passive bunkers supporting armed emplacements such as shelters, stores, searchlights, water supply, kitchens, medical centres, connected by trenches and tunnels.

London deep-level shelters. British public air raid shelters built under London’s Underground stations on the Northern line during WWII. Each shelter was a pair of parallel tunnels, with each tunnel divided into two decks and could house about 8,000 people, only eight shelters were built and they were not completed until 1942 after the end of the blitz. The shelters were opened back up in 1944 due to the V bombs sent by Germany.

See ‘air raid shelter’, ‘bunker’.

London defence positions. Late 19th-century CE scheme of fortifications in the southeast of England, designed to protect London from French invasion landing on the south coast.

Longphort, Longphuirt. (Ir.). A place on a river bank or river island where Vikings fortified as a base for raiding and trading.

Long walls of Athens. Parallel land walls connecting Athens with the port at Piraeus, later construction was started in 461 BCE and finished in 457 BCE, this provided Athens with a fortified link to the sea and supplies which meant that the city could not be besieged by land alone. Another land wall was added to the port of Phaleron in the 440’s BCE, to provide access to another port and would allow protected land for growing crops. The walls were still standing at the beginning of the 1st century BCE but the Romans laid siege to Athens and Piraeus (87–86 BCE) and won and then destroyed the Long Walls.

Long Wall of Quang Ngai. (Vietnamese. Trường lũy Quảng Ngãi). A defensive wall built by the Nguyễn dynasty, in 1819 CE, consisting of a rampart and small forts, 127.4 kilometre (79.2 miles) long, running from Quảng Ngãi Province in the north to Binh Dinh Province in the south, as a border with the H're people. Also known as the ‘*Great Wall of Vietnam*’.

Loop, loop hole, loup.

(1) A narrow opening in the wall or merlon of a fortification, through which missiles were discharged at the enemy, positioned to command the approaches, as well as protecting the weak spots. Designed to provide the maximum amount of protection for the defenders as well as give a reasonable field of fire.

See *'arrow loop', 'cheek', 'floor', 'gun loop', 'hooper loophole', 'throat'.*

(2) The circle at the end or at the centre of an arrow slit which was used to increase the field of fire. Also known as an *'oilet'.*

See *'arrow loop', 'arrow slit', 'oilet'.*

Loopholed walls and defended buildings. A British WWII defences using existing walls and buildings by adding embrasures, simple and camouflaged. A dairy storage in Beaulieu, Hampshire, England, was strengthened and windows were reduced inside to gun slits, while leaving the original windows on the exterior.

See 'British hardened field defences', 'FW3', 'pillbox'.

Lorica. The palisade made from *'villi'* or the sharpened stakes lashed together and rammed into the top of the ramparts forming the parapet surrounding a Roman encampment.

Low artillery outwork. See *'hornwork', 'ravelin'.*

Low wire entanglement. A low wire entanglement on wood pickets sunk into the ground with barbed wire only on the tops, where the wire can be hidden by low vegetation. Another version is where the pickets are flush with the ground and the wire is strung loosely, used in the Boer and Manchurian wars.

See *'barbed wire', 'entanglement', 'high wire entanglements', 'spider wire entanglements'.*

Lozenge pillbox. A British WWII pillbox which was built in the North East of England consisting of an irregular hexagon in plan, with long front and back walls compared to the sides, thus there is room for four forward facing embrasures, the rear wall was provided with two embrasures, an internal anti-ricochet walls runs along the long axis, defended by rifles and light machine guns. There are 176 remaining, one of them is located at Burnham on Crouch, England.

See 'British hardened field defences', 'FW3', 'pillbox'.

L-plan castle. A castle, keep or tower built with a ground plan in the shape of the letter *'L'*. Dating from between the 13th and 17th century CE, from Scotland to Romania.

See 'tower house', 'Z-plan tower house'.

Gleninagh castle, County Clare, Ireland.

Luftlandesperre (L-Spere). (G. *airborne barrier*). German WWII anti-glider landing obstacle.

See 'anti-glider trenches', 'Rommel's Asparagus'.

Luftschutzraum (LSR). (G. *air raid shelter*). A type German WWII air raid shelter which came in many forms, which usually include air tight entries, power plant, medical centre, water supply systems.

See 'flakturm'.

Lunette. (Fr. *lunette*, dim. of *lune*, moon).

(1) A small work generally situated before the curtain, in a ditch full of water, consisting of two faces forming a re-entering angle, and was used to enfilade the ditch.

(2) A field work consisting of two faces forming a salient angle and flanks parallel to the capital.

(3) A work situated on either side of the ravelin, where one face is perpendicular to half or two thirds of the faces of the ravelin, and the other face is the same size as the ravelin.

(4) A work sited forwards of the second ditch, opposite to the places of arms, the same as a ravelin except for its situation.

Lunette d'Arçon. A French small, detached fortlet dating from 1795 CE, triangular in form, defended by a ditch, reverse fire casemates under the counterscarp, guns in its face, a circular réduit tower in the gorge, provided with barracks, stores and magazines, looking like a 19th century CE *'polygonal fort'*.

See 'lunette'.

Luo cheng. (Ch. *rounding wall*). A defensive wall or extension of the outer city wall, a large wall city could have a number of '*lou cheng*'.

See '*cheng*'.

Photographs.

Acropolis, Southern slope, Athens, Greece.

Arrow embrasure (Pointed). *Guldursun-Kala in modern Karakalpakstan, Uzbekistan, was a fortress built in the 12th century CE.*

Artillery tower, dating from 1636 CE, Menton, France.

Barbican. *Caen Castle, France.*

Bartizan (inside view). *Fort George, Inverness, Scotland.*

Bartizan (external view). *Fort George, Inverness, Scotland.*

Bascule bridge. *Kalmar Castle in the province of Småland in Sweden.*

Bastion spur. *Fortezza Medicea in Siena, Italy, built between 1561 and 1563 CE.*

Battery tower. Vadstena Castle (Slott) - Sweden – outside.

Battery tower. Vadstena Castle (Slott) – Sweden, the inside.

Blind gun port. *Two blind gun ports on the square wall tower, Topakapi palace, South of main gate, Istanbul, Turkey.*

Boer War blockhouse, Wellington District, South Africa.

Box machicolation, *above a closed up gateway in the walls of St Catherine's Monastery, Sinai Peninsula of Egypt.*

Brattice. *Brattice on a tower of Casscasone, department of Aude, region of Occitania, France.*

Bridge head. *Kalmar Castle in the province of Småland in Sweden.*

Chateau. *Chateau Gaillard, Normandy, France.*

Cistern. *Internal cistern in the Crusader castle at Shobak, Jordan, with rainwater inflow channel on right.*

Coffer. *Agra fortress, Rajasthan, India.*

Cordon. *Auxonne, France.*

Corner buttress turret on a keep. *The White Tower, Tower of London, London, UK.*

Corner flanker, with arrow loops and gunports, *Legge Mount, in the north-western corner of the outer ward. Tower of London, England, UK, Dating between 1275 and 1285 CE. The gun ports were added in 1682-3 CE when the tower was heightened.*

Courtyard, keepless or quadrangular castle, *Bolton Castle, Yorkshire, England, UK, late 14th century (1399) CE.*

Crusader castle. *"Pharoah's" or "Coral" Island, Gulf of Aqaba, Egypt.*

Curvilinear wall. *Chateau Gaillard, Normandy, France.*

Dame. *Auxonne, France.*

Defence in depth. *Casscasonne, department of Aude, region of Occitania, France.*

Defence (Lines of). *The repaired (UNESCO grant - 1992) land walls of Constantinople, Istanbul, Turkey.*

Diatenshu. *Himeji castle, Japan.*

Ditch. *Maiden castle, Dorchester, United Kingdom, Multiple ditches and ramparts.*

Dobei. *Himeji Castle, Japan.*

Drawbridge. *Drawbridge of Fort George, Inverness, Scotland.*

Drum tower. *Drum towers of the Château d'Angers, city of Angers in the Loire Valley, in the département of Maine-et-Loire, in France.*

'D' shaped flanker, *dating to the Roman period, Carcassonne, France.*

'*D*' shaped flanker. 13th century CE, *Carcassonne, France.*

Dun. *Dun Telve, a broch at Glenelg, Scotland, from the outside.*

Dun *Telve, Glenelg, Scotland, from the inside.*

Embrasure. *Calshot Castle, Calshot Spit, Hampshire, UK.*

Embrasures (inside). Ajloun castle, Jordan.

Embrasure. *Marble tower, end of the Land walls of Constantinople, Embrasure in the interior and the exterior. The outer side of the embrasure in the 'Marble tower', Istanbul, Turkey.*

Embrasure. *Caen Castle, France.*

Enceinte. *Conwy wall, exterior of Mill Gate, in the southern wall, Wales, UK.*

Conwy wall, the south wall looking west, towers 21 to 18, including the Mill Gate, Wales, UK.

Forebuilding. *Rochester Castle, Kent, UK., stairs leading up to the entry to the forebuilding.*

Gate. *A gateway with gates covered in downwards overlapping metal plates 4-6mm thick, riveted to the gate, Soguk Çesme Sk., south west of the main gate of the Topkapi palace, Istanbul, Turkey.*

Gatehouse. *Harlech Castle, Gwynedd, Wales, UK.*

Gate tower. *Gate towers, Casscasonne, department of Aude, region of Occitania, France.*

Gateways. *Beumaris Castle, Anglesey, Wales, UK.*

Giri-durga. *Mehrangarh Fort, Jodhpur, Rajasthan, India.*

Grilles. *Nantes château, Nantes, Loire-Atlantique département, France.*

Gun port with slit. *Caen Castle, France, from the outside and from the inside.*

Hako hazama. *Himeji Castle, Japan, from inside.*

Hirayamajiro in the back ground with ishigaki and dobei walls in the foreground. *Himeji Castle, Japan.*

Hoard. *Caerphilly Castle, South Wales, UK, outside the hoard.*

Caerphilly Castle, South Wales, UK, inside the hoard.

Indian merlons. *Agra fortress, Rajasthan, India.*

Ishi-otoshi mado. *Himeji Castle, Japan, from below.*

Keep. *Rochester castle, Kent, England, UK, early 12th century (1127) CE.*

Keep, *Conisbrough Castle,Doncaster, UK, late 12th century (1190)* CE.

Kitchen. Krak des Chevaliers, Syria.

Kôrai mon. *Korai mon, Himeji Castle, Japan.*

Land walls of Constantinople. *Looking north from the Yedicule at the land walls of Constantinople, Istanbul, Turkey.*

Machicolations above a gateway. *Caen Castle, France.*

Machicolations atop of a tower, *Palace of the Grand Master, Hospitaller, Rhodes, Greece.*

Merlon arrow slit. *Merlon arrow slit in wall fronting the moat in front of the Golden Gate. Istanbul, Turkey.*

Moat. *Beaumaris Castle, Anglesey, Wales, UK.*

Moorish merlon. *Moorish merlons, Alhambra, Granada, Spain.*

Motte with a stone '*quarterfoil keep*' on top. *Clifford's Tower (a quatrefoil keep) on a motte, York, UK.*

Mural tower. *Half circle mural tower on the North side of the Yedicule tower complex, Istanbul, Turkey. Note: the separate platforms in front of the crenel as to the inner steps.*

Murder hole. *Murder hole at the top of the gateway, Trim Castle, County Meath, Ireland.*

Mushamada. *Himeji Castle, Japan.*

Nurhag. Su Nuraxi di Barumini, *outside the front of the megalithic fortress, Sardinia.*

Su Nuraxi di Barumini, *from the lowest access point of the tower, Sardinia.*

Su Nuraxi di Barumini, *from the lowest access point of the tower, Sardinia.*

Observation tower, *Fort Nepean, Victoria, Australia.*

Ogival bastion and points. *13th century CE Narbonne Gate, at the East entrance to the citadel of Carcassonne, France.*

Polygonal tower with a postern gate. *Between Yedikule and the Marble Tower with postern, Istanbul, Turkey.*

Postern. *At the base of the walls of "King David's tower", Jerusalem, Israel.*

Portcullis, *Chateau Gisor, Normandy, France.*

Pointed merlons. North west tower of the Yedicule, Istanbul, Turkey.

Put-log hole. *Left. Put-log hole inside the postern gate in the corner of the Polygonal tower between Yedicule and the Marble Tower. Right. Opposite side of the Put-log hole inside the postern gate in the corner of the Polygonal tower between Yedicule and the Marble Tower, Istanbul, Turkey.*

Ramparts. *The Ramparts of Fort George, Inverness, Scotland, UK.*

Redoubt. *Marie Therese Redoubte, Avrieux, France.*

Inside of a **sama** at Matsumoto Castle, Japan.

Sekizou no mon at Sogen ji Temple, Okinawa, Japan.

Shot deflecting battlements. *Caerphilly Castle, South Wales, UK.*

Slight. *Corfe Castle, Dorset, UK. Slighted using explosivees by order of Parliament following the English Civil Wars.*

Spur. *Chepstow Castle, Monmouthshire, Wales, UK.*

Square flanking tower. *Rochester Castle, Kent, UK.*

Square wall tower. *Square wall tower, Sea walls under repair 1992 as part of a UNESCO grant, Istanbul, Turkey.*

Talus. *Krac des Chevaliers, al-Husn, Syria.*

Theodosian (Land) Walls. *The Theodosian Land Walls, looking south to the Yedikule, Istanbul, Turkey.*

Tower House. *Dysert O'Dea Castle, Mollaneen, County Clare, Ireland.*

Trebuchet. *'Counterweight' trebuchet, Caerphilly castle, Wales, UK.*

Tunnel. Krak des Chevaliers, Syria.

Tunnels running through a rampart. *Auxonne, France.*

Turreted. *Corner turret, Rochester Castle, Kent, UK.*

Uchikomihagi. *Uchikomihagi ishigaki at Himeji Castle, Japan.*

Ward. *Beaumaris Castle, Wale, UK.*

Watari yagura. *Himeji Castle, Japan.*

Wicket *in the main gate of the Grand Master's Palace, Valletta, Malta.*

Yagura. *Tokyo imperial palace, Japan.*

Yagura mon. *Hishi gate, Himeji Castle, Japan.*

Yett. *Doune Castle, Perthshire, Scotland, UK.*

Zharakty gate. *Otrar, Kazakhstan.*

M.

Machiavelli, Niccolò (1469 – 1527 CE). Florentine author of the '*Treatise on the Art of War*', published in 1521 CE.

Machicolated. Provided with machicolations.

Machicolated battlements. Battlements provided with machicolations.

Machicolated corner gallery. A wall gallery situated at the corner of a fortification; usually provided with arrow slits, which added to the defensive capacity of the wall.

Machicolated gallery. A gallery provided with machicolations, situated at the top of a curtain wall or just below the level of the wall walk, dating from the 14th century CE.

Machicolated gatehouse. A gatehouse provided with machicolations.

See '*gatehouse*'.

Machicolated turret. A turret which was built out on corbels, the holes between the corbels were used as machicolations. Also known as a bartizan or machicoulis.

See '*flanking turret*'.

Machicolated wall walk. See '*machicolated battlements*'.

Machicolation. (Fr. *macher*, to crush). A series of openings provided by. building the parapet out on corbels, projecting beyond the face of the wall, the space between the corbels providing the openings for the machicolations; or by leaving gaps between a recessed wall and the buttressed arch standing before it. Projectiles and liquids could be thrown onto the enemy at the base of the walls, thus reducing the dead ground.

The term is usually reserved for the stone constructions which in many countries superseded the wooden hoardings used for the same purpose about the middle of the 14th century CE. They were used to defend the walls and towers of a castle, and were used specifically for the defence of gateways and other entrances, which were the weakest part of a fortification. By the 15th century CE, the arch supporting the machicolation was replaced by a plain lintel.

See '*Briton machicolation*', '*Italian machicolation*', '*machicoulis*'.

Machicoulis.

(1) A single or a small number of machicolations, positioned over a gate, doorway or some other vulnerable angle of the defences. Developed as a substitute for the hoard (as did the machicolation) but served to provide particular defence of a position an example is the box machicolation.

See *'box machicolation'*.

(2) Machicoulis were also built around the angles of towers and curtain walls, often semi-circular in shape, known as bartizans, which were used to defend the area below as well as flanking the neighbouring positions.

See *'bartizan'*, *'flanking turret'*.

(3) A square aperture in the floor of the room above an entry, which was used to attack the enemy from above, usually covered by a trap door, and was known as a '*murder hole*'.

***See* 'murder hole'.**

(4) Machicoulis has also been used to refer to machicolation in some texts.

Machine gun nest. A position occupied by a crew of two or three soldiers armed with a machine gun.

Mafumi. (J.). The flat top of an earthen embankment ('*doi*') of a castle, which usually included a fence or wall ('*inubashiri*'). Also known as '*bafumi*' or '*hirami*'.

See *'dote'*.

Maginot line. A fortified area of the French border, built in the 1930s CE to deter invasion by Germany, using various artillery and gun emplacements to defend the borders with Italy, Switzerland, Germany, Luxembourg and Belgium. Named after the French Minister of War André Maginot.

Components of the Maginot line included; anti-tank obstacles, anti-personnel obstacle (barbed wire entanglements), artillery, anti-tank-gun and machine-gun emplacements, barricades, border guard posts, communications centres, infantry shelters, observation posts, supported by the natural terrain of dense forests, rivers and steep impassable topography.

The emplacements contained a range of infrastructure such as; ammunition stores, barracks, kitchens and mess halls, hoists, medical centres, power stations, storage for spare parts and food, ventilating systems, water storage and distribution systems, and workshops.

The emplacements were connected via a network of tunnels, including a narrow gauge electric railway for transport between bunker systems.

Ultimately in WWII the Germans bypassed by going through Belgium, thus showing up the flaw in the use of static line defence and a waste of public money.

***See* 'abri d'intervalle', 'anti-tank rail', 'bloc', 'cloche', 'Commission d'Organisation des Régions Fortifiés (CORF)', *'dragons teeth'*,**

'eclipsing turret', 'entrée', *'first fortified belt', 'fort palmé', 'fossé diamant', 'goulette à grendage', 'linear fortifications', 'tunnels', 'turret'.*

Magiri. (J. *section breaks*). A way of building a castle's wall (*'hei'*) in disconnected sections, so if a section of wall was damaged during an attack the adjoining section would remain intact.

Magistral line. The main line of defence or the crest of the scarp of a fortification. In drawings of fortifications it is usually drawn in a thicker line than the other lines.

Maham tower. A siege tower constructed of logs, earth and *'gabions'* used during a siege to provide a tower so that the besieger could fire over the defender's parapets. Used during the siege of Fort Watson during the American Revolutionary War, and is named after Lt Col. Hezekiah Maham.

Mahi-durga. (Sanskrit). An Indian mud or earthen fort that came in three varieties; see *'mrid durga'*, *'parigha durga'*, *'panka durga'*.

See 'durga'.

Main d'Oeuvre Militaire (MOM). (Fr. Military Labor). The French engineering corps built various concrete works during the 2nd phase of the *'Maginot Line'* between 1935 to 1940 CE, consisting of light bunkers and casemates.

See 'Maginot Line'.

Main guard. The strongest building of a fortification, usually the *'keep'*.

Main line of defence. A line along which the major strength and effort of defences is constructed.

Main line of resistance (MLR). The most important part of an army's defensive position facing the enemy which will consist of various fighting positions, trenches, pill boxes, etc.

See *'London defence positions'*.

Main operating base (MOB). An American permanently manned and well defended base, with sea and/or air access, to support *'forward operating bases'*, used to address regional threats in Africa, Asia and South America.

See 'forward operating base'.

Main work, Major work. The chief work of a fortification, as opposed to the outworks.

Maghribi. A large Indian catapult. **See *'manjaniq'*.**

Malloppo. (It. *swag*). A loopholed position in either a cave or casemated battery, linked to other parts of the work by passages, part of the Italian Alpine Wall, post WWI.

Malvoisin, Malregard.

(1) A type of siege tower.
(2) An earthwork fort or mound designed to be used to blockade the approaches to a fortification under siege. An example of use was by Edward 1 at the siege of Stirling in 1304 CE.

See *'counter castle'*, *'gegenburg'*, *'tai no shiro'*.

Mamelon. (Fr. *nipple*). A breast shaped hillock hastily fortified with trenches, re-enforced with timbers, gabions and sandbags, used during the Crimean War, 1854 - 1855 CE.

Ma mian. (Ch. *horse face*). A rectangular bastion. Also known as *'Zhi'* and *'Ying lou '* (Ch. *hard tower*).

Man cheng. (Ch. *Manchu city*). ***See 'man zhou cheng'.***

Manidwar. (Sanskrit). A window in the gate of a fort.

See 'grille', 'speculartorium'.

Manga, Mangana. Diminutive form of mangonel.

See 'mangonel'.

Manjaniq, Maghribi. (Ar.). Basically the staff-sling arm mounted on a frame, a catapult.

See *'trebuchet'*.

Mangonel. (Ar. *Al-majanech*, crusher or mangler), (Gr. *Mangono*, to crush or squeeze). A catapult consisting of a wooden arm with a large spoon or basket on the end, attached to a wheeled frame, with its bottom end surrounded by rope or rawhide. The rope was attached to a windlass which was turned by wooden levers creating tension on the arm. When this tension was released the arm came to stop when hitting a cross beam, thus the contents would be flung forward; mangonels were used at the siege of Syracuse in 350 BCE.

Mannheim Line. A fortified line built by Finland across the Karelian Isthmus (140km (87 miles)) against Russian attack, built in two phases (1920–1924 and 1932–1939 CE), supplied with ditches, barbed wire entanglements and pill boxes.

Manor house. See *'fortified manor*'.

Mantelet. A large Roman shield made of wood, carried by two soldiers for protection while assaulting a fortification.

Mantle. *See 'turret'.*

Mantlet wall. (L. *mantellum*, cloak).The outer wall of a late medieval castle that was situated between the main curtain and the moat.

Mantrap. ***See 'booby trap', 'mine', 'punji sticks', 'trip wire'.***

Man zhou cheng. (Ch. *Manchu city*). A walled city within a city erected during the Qing period to act as a garrison, and separate the Manchus from the Han Chinese. Also known as either '*Man cheng*' (*Manchu city*), '*Qi Cheng*' (*Banner city*), '*Zhu fang cheng*' (*Garrison city*).

See '*cheng*'.

Manzil. (Ar. *house or dwelling*). A fortified hostel on the Iberian Peninsula.

Marching Camp. A Roman '*marching camp*' was a temporary fortification set up to protect the troops while marching through hostile territory, consisting of a surrounding ditch and rampart, with a palisade.

See '*castra*'.

Mareth Line. A French line of defence built between Medenine and Gabès in southern Tunisia 40km (25 miles) to defend against an attack by Italy. After the fall of France in WWII the line was used against attacks by the Allies, which was re-enforced by casemates, bunkers, mines, barbed wire and anti-tank obstacles.

Bunkers on the Mareth Line, Tunisia.

Marsad. Small isolated fortification along the main roads for military traffic of Andalusia, Spain. Also known as a *'rutba'*.

Martello towers. English (1805 to 1808 CE). Small keeps, conical in cross section, oval in plan, with smallest section facing the sea, 4 metre (13 feet) thick walls, solid roof against plunging fire, surrounded by a ditch, entrance via upper floor by ladder or drawbridge. Provided with a 24 pound cannon able to be traversed a full 360 degrees on a rotating platform on the roof. Garrisoned with up to 24 soldiers and officers, provided with store rooms for ammunition. Pressed back into service after the fall of France in 1940 CE as part of coastal defences.

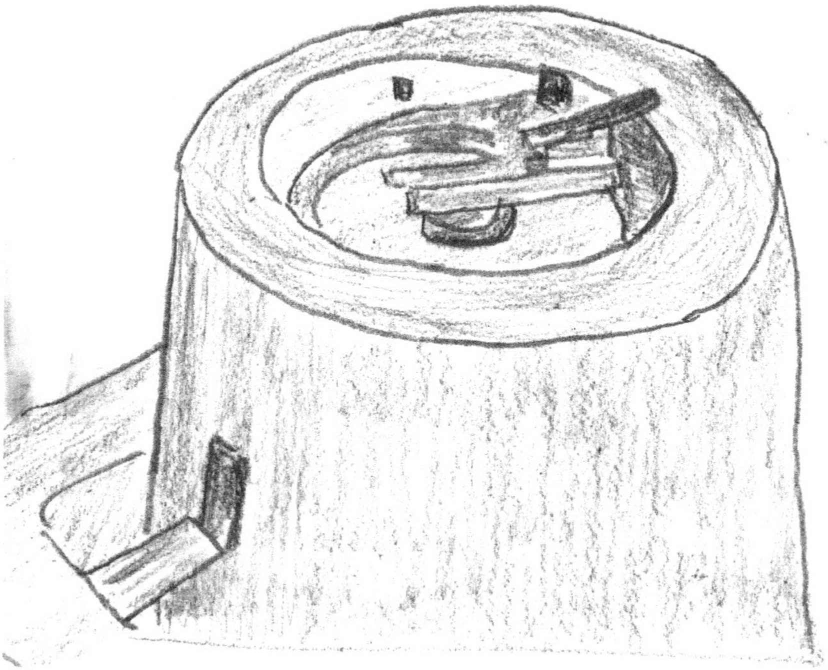

Dymchurch Martello Tower, Kent, England.

Maru. (J.) A bailey in a Japanese castle in the Edo period. Also known as a *'kuruwa'*.

See *'kuruwa'*.

Maru-durga. (Sanskrit, *desert fort*). A fort surrounded by an arid zone of at least 73 kilometres (45 miles). Also known as *'dhanu durg'*, *'dhanva'* *'dhanvana'*, *'nirudaka durga'* (*waterless fort*).

Maru umadashi. (J.). A semi-circular earthen embankment or wall built in front of Japanese castle's gate.

See *'barbican'*, *'doi'*, *'hori'*, 'kaku umadashi', *'koguchi'*, *'umadashi'*.

Mastio. An Italian term meaning donjon or keep.

Masugata. (J. *square measuring box*). A square surrounded by earthen embankments, stone walls, moat and gates, used to protect a castle's entrance (*'koguchi'*). Entry to the masugata was via a barrier (*'umadashi'*) and a small gate (*'korai mon'*). A right hand turn was needed to enter the next gate (*'ootemon'*). Used to muster troops before a sortie or act as a killing ground if the first gate was breached. Masugata could be constructed inside the main entrance (*'uchimasgata'*) or outside the main entrance (*'sotomasugata'*).

See 'koguchi', 'korai mon', 'masugata mon', 'ootemon'.

Masugata mon. (J. *measuring box gateway*). The barbican gatehouse used to defend the main access route of a Japanese castle, positioned at either the crossings of the inner, middle, or outer moats, or at the main entrance to the castle. The barbican gatehouse consisted of a walled square or rectangular courtyard provided with two gateways. The initial entrance was defended by a *'kôria mon'* (a small gate provided with a central roof flanked by two smaller ones; the second gateway was a *'watari yagura mon'*, a large two-storey gatehouse).

The two gateways were set at right angles to each other forming a walled courtyard, so if any enemy force gained entry to the courtyard they would be confined within that space, and have to turn to attack the formidable *'watari yagura mon'*, whilst under attack from the castle defenders. The small size of the *'kôria mon'* restricted the number of enemy soldiers entering the courtyard, thus the defenders had a greater chance of defending the *'masugata mon'*.

See 'koraimon', *'mon', 'yaguramon'*.

Masugata koguchi. (J.). A type of castle entrance consisting of two gates at right angles with a square box in between, surrounded by earthen embankments or stone walls. The outer gate (*'ichi no mon'*) was usually smaller than the inner gate (*'ni no mon'*).

See *'koguchi'*, *'masugata'*, *'mon'*.

Maunsel forts. English anti-aircraft forts (Army/Navy forts) dating from 1940 to 1944 CE, built in the Thames and Mersey estuaries to protect shipping. Guns situated on top of magazines, provided with an engine room, accommodation, radar and radio infrastructure. Built on top of concrete cylinder legs on a raft that was sunk and positioned on the sea bed in groups, connected by walkways. Designed by G.A. Maunsel, a civil engineer.

Maximilian Towers. Part of the Prussian system used by the Austrians at Linz (1830-1 CE), a chain of thirty two isolated towers, which were sunk into the ground within a ditch, with three floors masked by a glacis. Eleven 24 pounder cannons were on the top of each tower, arranged to graze the glacis.

See *'Mortello towers'*.

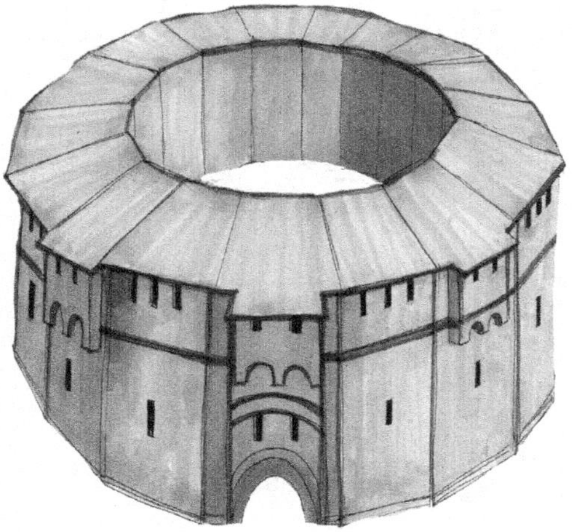

Maximum flanking effect. The much sought-after ideal amount of flanking effect, which increases a fortification's defensibility.
Mechanè. (Gr.). Siege engine.
Median wall. A wall built by the Babylonians in about 800 BCE between the Euphates and Tigris rivers, 80 km (50 miles) long, to repulse the attacks of the Medes.
Mediterranean Wall. *See 'südwall'*.
Melinite. The French adopted melinite for high explosives, while the British used '*lyddite*' and the Germans used '*guncotton*'.
Merlon. (I. *merlone*, battlement).
(1) The portion of a battlemented parapet that rises up from a wall (e.g. the solid part of a parapet between the crenels), which provided some protection to the defenders behind them.
(2) The part of a parapet between two embrasures.
See *'battlement', 'crenel', 'embrasure', 'guelph merlon', 'half merlon', 'Irish battlement', 'parapet', 'stepped merlon', 'Venetian battlement'*.

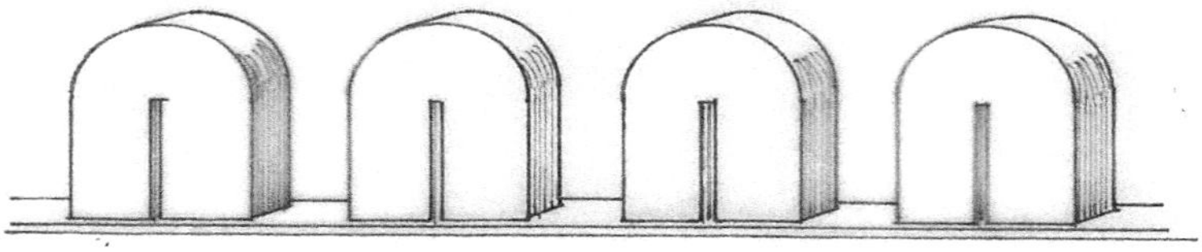

Arabic, Egyptian.

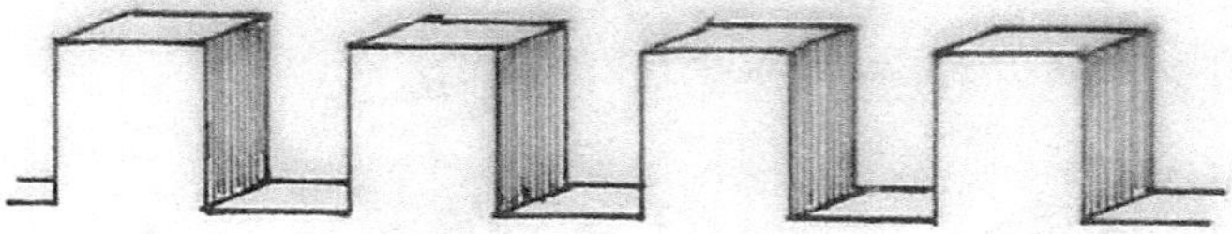

Common.

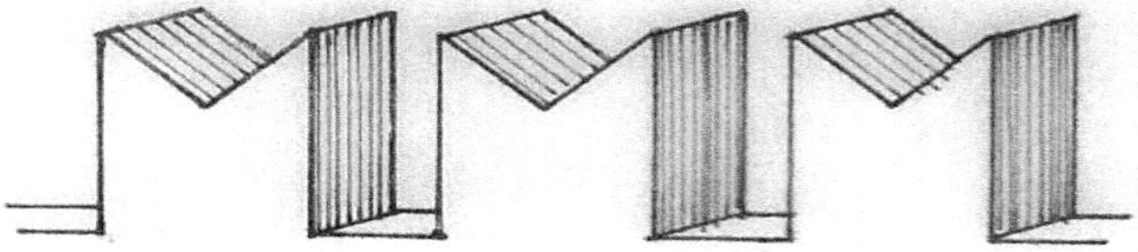

Guelph.

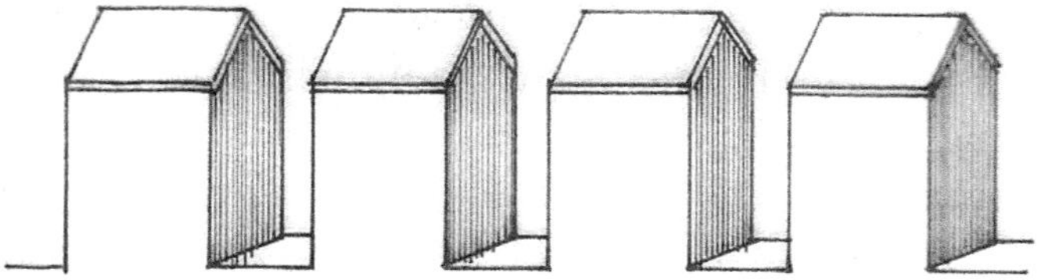

Pointed (A).

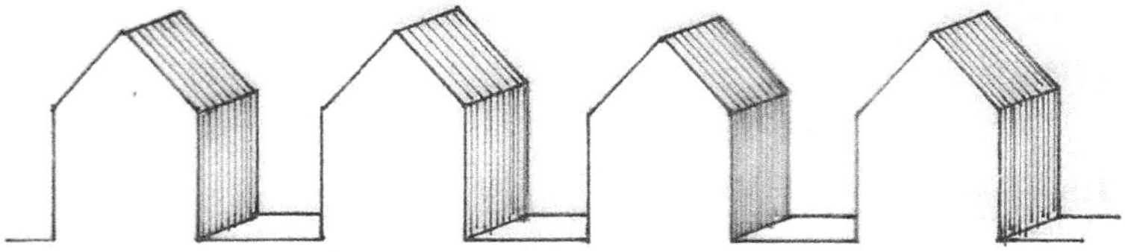

Pointed (b).

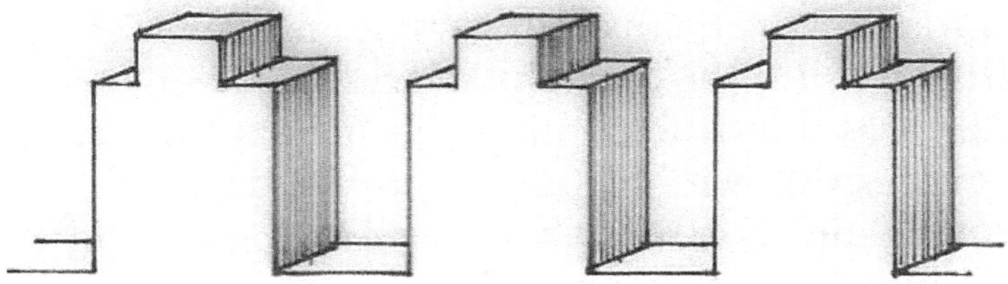

Stepped.

Merlon arrow slit. An arrow slit which was situated in the centre of a merlon, the addition of the arrow slit added to the potential fire power of the wall head defences. They also provided better protection to the archers than firing through the crenels.

See '*arrow slit*', '*merlon*'.

Metaxas Line. (Gr. Γραμμή Μεταξά, Grammi Metaxa). A line of 21 independent forts (from Popotlivitsa to Nymfaia) constructed by Greece along the Bulgarian border (155km, 96 miles) built between 1936 to 1941 CE but was only in use till 1945 CE.

Meurtriéres.

(1) Small loops through which the enemy could be attacked, usually situated in the gatehouse. Also known as murder holes and spy holes.

See '*drop box*', '*fall trap*'.

(2) Small gun loops through which muskets could be fired at the enemy.

See '*musket port*'.

Mezzaluna. (It. crescent moon). A demilune.

***See* '*demilune*'.**

Michelmannstand. (G. *Michelmann bunker*). A small German WWII prefabricated concrete bunker in the shape of a hexagon, with an open weapons emplacement and provided some shelter for the soldiers.

***See* ''*tobruk*'.**

Middle bailey. A bailey which was situated between the outer and the inner baileys of a castle.

See '*bailey*', '*concentric castle*'.

Migdol. (Hebrew, *tower*). Ancient Egyptian small forts built at oases and cross roads to control movement of hostile forces until the standing army could arrive.

Mikazukibori. (J.). A curved moat on the outside of the '*maru umadashi*' in front of a castle's main gate ('*koguchi*').

See '*bori*', '*doi*', '*koguchi*', '*maru umadashi*', '*umadashi*'.

Mile castle. A Roman fort which was used to defend a part of a fortified perimeter (e.g. Hadrian's Wall) situated every Roman mile along the length of the perimeter, (approximately every 1.5 kilometres. 0.9 miles). A mile castle was constructed on a rectangular plan covering 300 to 400 square metres (3229 to 4306 feet), provided with a barrack-room, a store, an oven and a latrine. There were two gates; the inner leading to civilized lands while the outer was provided with turrets, and opened to the barbarian lands. Also known as a '*castellum*'.

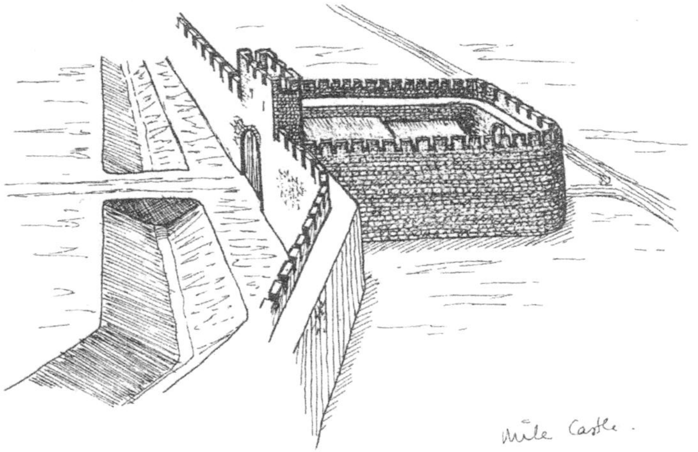

See 'Appendix 1. What is a Castle?'.

Military crest. The highest part on the slope on a hill which allows defenders to see and fire upon the base of the hill.

Military Way. A Roman road that ran behind a Roman frontier (e.g. Hadrian's wall) and north of the vallum, and linked the turrets, milecastles, and forts.

Mine.

(1) The works excavated by the besiegers to attack a fortifications defences by making a breach. The early method was to fill the excavated cavity under a wall with combustibles and set it alight. Once the upright timbers burnt through the chamber collapsed bringing down the wall above, thus making a breach. The later method of using a mine to make a breach involved using gunpowder, the gunpowder was

exploded in the excavated chamber under the walls, the resulting effects producing a breach.
(2) Mines were made by besiegers while countermines were made by the besieged.

See 'ecoutes'.

(3) Anti-personnel/vehicle explosive device laid to defend large areas of sea or land (i.e. Claymore mine). Types of mines include:

A) Anti-personnel mines; fragmentation and stake mines, shaped charge mines, directional mines, blast mines, bounding mines, flame mines, chemical mines.
B) Anti-vehicle mines; blast mines, shaped charge, full width mines, side attack mines, wide area mines, anti-helicopter mines.

See *'demining'*.

See Appendix 3. Siege Warfare - The art of Offence and Defence.

Mine chamber. The chamber at the end of an excavated mine under the intended target, were either the combustibles were ignited to burn out the wooden props or the explosives were detonated.

See 'mines (1)'.

Mine detection.

(1)The process of mine digging could be detected by placing a drum with a small stone on the skin and mine works would make the stone vibrate. By moving the drum around and 'listening' for the vibrations of the digging. A counter mine could then be dug and an underground battle would happen.
(2)The process of detecting anti vehicle and personal mines with equipment or trained animals such as rats.

Minefield. A large area using anti-personal/tank explosive devices laid to defend access routes of sea or land.

See *'mine 3)'*.

Minefield (Controlled). A minefield observed and controlled from a central point, part of the armed coastal defence, UK in the 1880's CE.

Minefield control tower. Two story observation tower made of concrete, located at Holliwell farm, Burnham-on-Crouch, Essex, England, built in 1940 – 41 CE, part of Essex coast defences. 6.7 m (22ft) high, 6m (20 ft) wide, hexagonal shaped, pill box for defence on lower level, upper level supplied with observational slits / embrasures. Overlooking the surrounding minefields and across the mouth of the River Crouch.

Minensperre. (G. *mine field*).

See 'mine field'.

Minor work. A lesser work which was used to defend a position or a major work.

See *'outwork', 'main work'.*

Missile silo. A reinforced concrete bunker in the shape of a vertical cylinder topped by a blast door, connected to a *'missile launch control centre'*, to house and launch a ballistic missile. Also known as a *'missile launch facility'* or more simply *'launch facility'* (LF).

Mirador. (Sp. *to look out*). A small lookout tower on top of a building, later a surveillance post for military camps and bases.

Mitre. See *'priest cap'.*

Miyagura. (J.). Watchtower.

See *'yagura'.*

Mizu no guruwa. (J.). A water compound or bailey of a castle.

See *'kuruwa'.*

Mizu no te guruwa. (J.). A type of castle compound (*'kuruwa'*) constructed to protect the castle's water, be it a well or reservoir.

See *'kuruwa'.*

Mizute kuruwa. (J.). A small bailey housing a well or reservoir.

See *'kuruwa'.*

Mizuki. (J. *water fort*). A large earthen embankment which was used as a dam to create a water defence against an attacking force.

Moat. (Fr. *motte*, mound). A water filled ditch which surrounded a castle, or just forming a line of defence, which was artificial or partly natural in origin. Just as 'donjon' signified the highest part of a motte and bailey castle and was later changed to *'dungeon'*, so moat was derived from *'motte'* or the mound to the ditch from which it was excavated. In some cases, the word moat is used to refer to a dry ditch. A moat filled with water significantly impeded the tactic of mining, which involved digging tunnels under a castle to collapse its defences.

See *'ditch'.*

Moated. Provided with a moat.

Moat pier. A pier of rock set amid a moat or ditch which was used to support a wooden bridge or drawbridge. Common in the Middle East because of the lack of strong enough timber to span the whole moat. Some piers were left standing when the moat was hewn out of the solid rock (an example can be found at Sahyun Castle, North West Syria). Also known as *'drawbridge pier'*.

Mock machicolation.

See *'blind machicolation', 'bracket', 'false machicolation'.*

Moesian Limes. (L. *Limes Moesicus*). The defence line consisting of legionary fortresses, forts and watch towers linked by a road (588km 365miles, *needs verification*), built by the Romans in the province of Moesia (Hungary), along the Danube river, dating from the 1st century CE.

See 'Antonine wall', 'foss', 'Hadrian's wall', 'limes', 'limes Raetian'.

Moineau, Moyeneau. (Fr. *moineau*, sparrow). A small flat bastion, usually situated in the middle of an overlong curtain, from which musketeers defended the neighbouring bastions with small shot flanking fire, otherwise the bastions would be inadequately defended due to their distance apart.

Moir pillbox. A British WWI prefabricated concrete pill box armed with a Vickers machine gun, which had a shutter for the gun but no other method of removing the fumes gun, built in Belgium and France in 1918 CE.

See 'pillbox'.

Molotov Line. (Russian Линия Молотова, Liniya Molotova). A fortified line built by the Soviet Union between 1940 and 1941 CE, along the western borders, the Baltic Sea to the Carpathian Mountains (100km, 365 miles), consisting of mutually supportive groups of bunkers armed with machine guns, anti tank guns and artillery.

Mon. (J. *gate*). A gateway of a Japanese castle, which was of two main types; the gateway which was constructed of timber similar in style to a '*tenshu*', and the other gateway was simpler and built into the stone walls or the '*ishigaki*' of a castle.

See '*kôrai mon*', 'masugata mon', 'munamon', '*uzume mon*', '*watari yagura mon*', '*yagura mon*'.

Moncrief mounting. A counterweight system used on a gun to help it disappear after firing.

Monk. A masonry obstacle built on top of a batardeau.

See 'batardeau', 'cofferdam', 'dame'.

Monomiguruwa. (J.). A compound built on high ground for a watch tower.

Monomiyagura. (J.). A watch tower located in the main compound ('*honmaru*') of a castle.

See '*yagura*'.

Montalembert, Marc-René. (1714 – 1800 CE). The Marquis de Montalembert, French general and military engineer, responsible for replacing star fortresses with simpler polygonal fortresses, which became the standard system in the early 19th century CE.

Monte-charge. (Fr. *freight elevator*). The goods lift in French late 19th century CE forts and the *'Maginot Line'* used to transport ammunition to the gun emplacements above.

See 'Maginot Line'.

Moon. An outwork constructed in the form of a crescent.

See *'demilune', 'mezzaluna'. 'ravelin'.*

Moorish merlon. A merlon which was typical of the fortified *'Mudéjar'* architecture, characterized by a pointed apex on all four sides.

See *'merlon'.*

Morice Line. A long obstacle built and designed by the French in 1957 CE to stop the infiltration by guerrillas from Tunisia and Morocco to Algeria, 460km (286 miles) long along the border with Tunisia and 700km (435 miles) long along the border with Morocco (totalling 1160km, 719miles). Consisting of electrified barbed wire (5000 volts), with alarms, radar and searchlights, inside a mine field, patrolled by troops and supported by artillery, helicopters and tanks.

Morrison shelter. A UK kit form bomb shelter during WWII, consisting of a cage like structure made from steel and welded mesh, measuring 1.98m X 1.2m X 0.76m (6 feet 6 inches X 3 feet 3 inches X 2 feet 6 inches).

See *'Anderson hut', 'flak towers (Germany)', 'Stanton air raid shelter', 'Winkel bunker'.*

Mota. (Fr.). ***See 'motte'.***

Mote, Motte. A natural, partly natural or artificial earthen mound surmounted by a wooden donjon surrounded by a palisade, and in turn surrounded by palisaded bailey or the bailey was placed to one side; the motte was usually surrounded by a ditch which was also surrounded the palisaded bailey. The motte and bailey were connected by a flying bridge. The *'motte and bailey castle'* was common in France and was introduced to England by the Normans, who used them to consolidate their territorial gains. Square motte and bailey castles also existed and there is the example of Dundaff Castle, which lies in the Fintry hills just outside Denny, Scotland. Danish mottes also came with square outlines.

See *'kernmotte'.*

See 'Appendix 1. What is a Castle?'

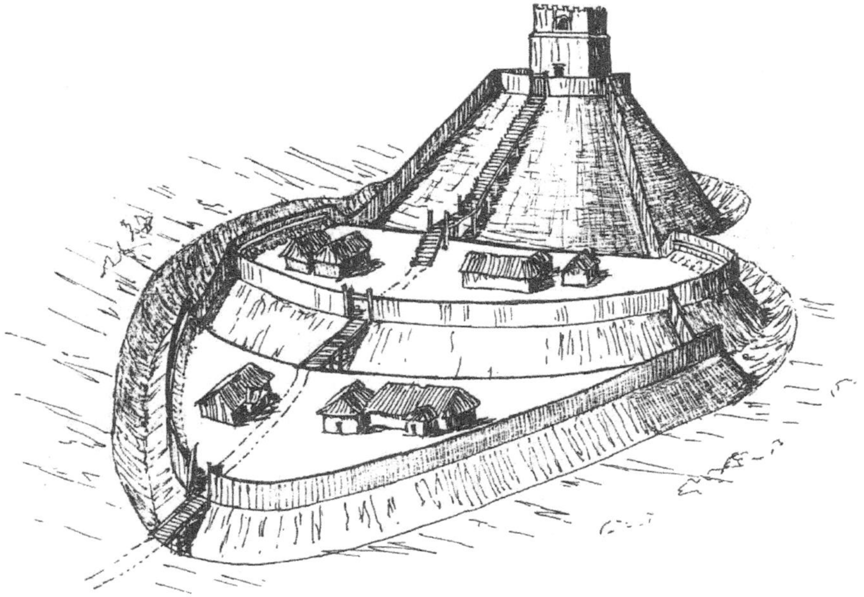

Motte and Bailey castle, consisting of a motte and a double bailey, Brinklow, Warwick, UK.

Motte and bailey castle. See *'bailey'*, *'motte'*.

Motilla. A fortified settlement around a well, dating from the Bronze Age on the Iberian Peninsula, consisting of a tower with three lines of concentric walls, a large courtyard and a well. An example is Motilla del Azuer, near Daimiel, province of Ciudad Real, Spain.

See 'nurhag'.

Mougin turret. (Fr.). A land based revolving gun turret used on French fortifications in the late 19th and early 20th century CE, including the *Séré de Rivières system*' fortifications in the 1870's. The turret is fitted with two 155 mm guns under an armoured shield, sunk into the ground and surrounded by a thick concrete apron.

See *'gruson and shuman turrets'*, *'Séré de Rivières system'*.

Mouse.

(1) A wooden penthouse, with a gabled roof covered in animal hides, moved forward on rollers, so sappers could attack the base of a fortification's wall. Also known as a '*musculus*'.

(2) A metal tipped wooden lever used to pick out and remove stones from a wall of a castle to make a breach in a wall. Could be used under a penthouse for the protection from falling objects.

See *'bore'*.

Mouth. The narrow opening of an embrasure, at the interior slope of a parapet.

See '*embrasure*'.

Mouton. (Langue d'oil). A battering ram.

Mrid-durga. (Sanscrit). An Indian fort surrounded by earthen walls.

See 'durga'.

Mudéjar merlon. A merlon which was built with a pointed apex, typical of *Mudéjar* fortified architecture.

See '*moorish merlon*', '*pointed merlon*'.

Multiple lines. A number of lines of detached walls which were used to defend a position.

See '*concentric castle*', '*lines of defence (1)*'.

Multivallate hillfort. A hillfort which was surrounded by a number of ditches and ramparts, which were either built originally as a multivallate hillfort or had a number of ditches and ramparts added, thus a univallate or a bivallate hillfort could develop into a multivallate hillfort. This development in hillfort design was made necessary by the importation of the sling to Britain, the sling which had a longer range compared to that of the throwing spear. This made it necessary to keep the enemy further away from the principle defences, this was achieved by extending the outworks. This however, does not suggest that all univallate hillforts pre-date all multivallate hillforts, an example in Maiden Castle, UK.

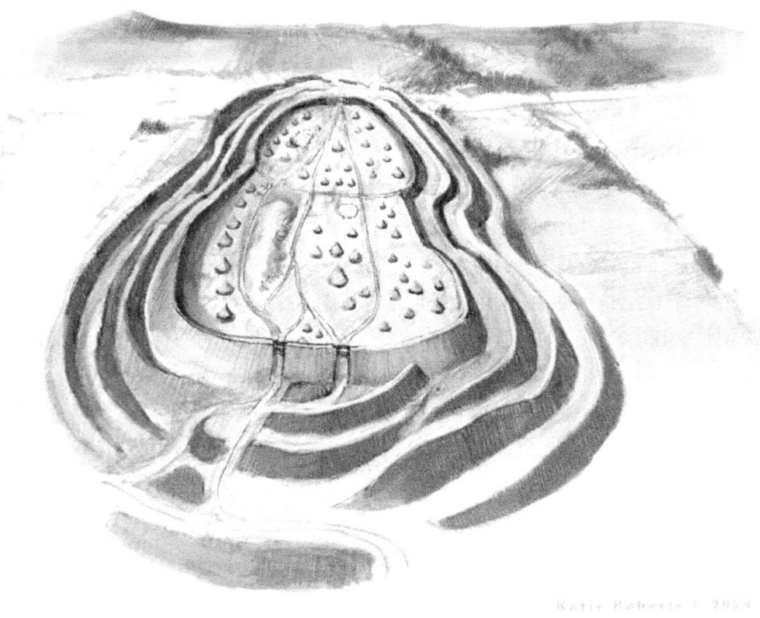

Maiden castle, Dorchester, UK.

See '*hill fort*'.

Mu man. (Ch.). A mobile protective screen.

See '*bu man*'.

Munamon. (J.). A gate consisting of two main pillars and a roof, similar to a '*koraimon*', usually supported by earthen or stone walls.

See '*koraimon*', '*mon*'.

Mündungsgasgrube. (G. *muzzle gas pit*). A large inclined concrete slab under an embrasure in some German WWII emplacements which was intended to disperse the fumes from the firing guns.

Munitio. (L. *fortification*).

Mura no shiro. (J.). Village castle. **See '*shiro*'.**

Mural gallery. A gallery constructed within the thickness of the wall below the parapet, which was provided with arrow slits.

See '*firing gallery*'.

Mural tower. (L. *murus*, wall). A tower attached to the curtain wall which was used for flanking the curtain wall.

See '*bastion*', '*curtain tower*', '*wall tower*'.

Murderes. The square gun ports which were used in coastal artillery forts of the late 15th and early 16th century CE, England.

Murderess. The gap of a machicolation, which was used to attack the enemy at the base of the wall below.

See *'machicolation'*, *'murder hole'*.

Murder hole.

(1) An opening in the floor of the room of a gatehouse situated above the entry passage or in other important places, which was used to attack the enemy below.

See 'drop box', 'machicoulis', 'meurtriéres'.

(2) The gaps of the machicolations which were used to attack the enemy at the base of a castle's walls.

Examples of murder holes can be found in: Byzantium, Constantinople, Land wall gates, 600's CE; Narbonne gate at Carcassone, France, 13th century CE; Great gatehouse at Harlech, Wales, UK, 1285 CE; Deal Castle gateway, Kent, England, 16th century CE.

See *'machicolation'*, *'murderess'*.

Murrette. (Fr. *small wall*). A small stone wall used by the French mainly in north Africa when the ground was too hard to dig a trench for camps or combat positions during the 19th and 20th centuries CE.

Murus. (L. *wall*). A defensive wall built around a Roman town.

Murus caespiticius. (L. turf wall). A Roman fortified wall made from turf.

See 'Antonine wall'.

Murus Dacicus. (L. *Dacian wall*). The Roman defensive walls constructed in Dacia.

See 'Antonine wall', 'Hadrian's wall', 'limes', 'limes Raetian'.

Murus gallicus. (L. *Gallic wall*). A way of constructing the walls of Iron Age hillforts or *'oppida'* using inter-connected timbers as a matrix which was filled with dirt and stones, with an outer stone facing of the ramparts.

See *'crib work'*, *'intra-mural timbers'*, *'timber laced rampart'*.

Musculus. (L. *muscle*). **See *'mouse'*.**

Mushabashiri. (J. *warrior run*). Part of an earthen embankment (*'doi'*) or stone wall (*'ishigaki'*) used to move soldiers about a castle. Also known as *'ausaka'* or *'aisaka'*.

See *'kasanezaka'*.

Musha gaeshi. (J. *warrior return*). Ishigaki where the stone is built is such a way to deter climbing by an adversary. At Naruse castle (Toyota, Aichi Prefecture), the top of the *ishigaki* gently curves back over the wall below like a wave breaking. Hitoyoshi castle (Hitoyoshi, Kumamoto Prefecture) the top of the *ishigaki* is topped by rectangular stones projecting out over the main wall making a barrier to climbers.

See *'ishigaki'*.

Mushamada. (J. *warrior window*). A window with 3 or 5 vertical lattices, used to fire/loose missiles through. Also known as a *'tategoushimado'* (J. *vertica lattice window*).

Mushrabiya. (Ar.). A machicolated balcony, sometimes provided with parapet, hanging over a gateway.

Mushroom pillbox. A British WWII concrete pillbox designed and built in 1940 CE, circular in plan view, with an overhanging concrete dome on the roof which offered great views but little protection to the occupants from direct fire, common in airfields of southern England.

See 'pillbox'.

Musket port. A gun port which was designed for the use of muskets, dating from the 16th century CE. Also called *'musket embrasures'*.

See *'gun port'*.

Mutual support. Two fortifications or parts of the fortification within the *'field of fire'* which provide fire to protect one another.

See 'arc of fire'.

N.

Nachrichtenstände. (G.). Communications post.

Nagadote. (J.). A long earthen defensive embankment raised outside a castle.

See *'doi'*, *'dote'*, *'kazashi no dote'*.

Nagayamon. (J.). A gate way through a long warehouse, accommodation, or as a guard house around samurai homes.

See *'mon'*.

Nähkampfraum. (G. *combat room*). A German WWII small caponier at the rear of a bunker to defend the entry, with a port hole with sliding metal shutter.

See 'caponier'.

Nahkampfwerk. (G. *Melee work*). Austrian infantry position in advance of the artillery created as part of a new type of strongpoint system in 1900 CE.

Nakajou. (J.). The second compound of a Japanese castle, usually housing the lord's close family and retainers. Also known as a '*ni no maru*', '*ni no kuruwa*', or a '*naka no maru*'.

See *'honmaru'*.

Nara-durg, Nri-durga. (Sanscrit). An Indian '*human fort*', a fort defended by a large number of loyal warriors, usually a city based fortress provided with a large garrison. Also known as a '*bala durga*'.

See 'durga'.

National Redoubt. (Fr. *Réduit national*).

(1) An area to which the (remnant) a country's forces can be withdrawn if the main battle has been lost or even earlier if defeat is considered inevitable, could be mountainous or a peninsula.

(2) Belgium's National Redoubt was based on Fortress Antwerp with its double defensive belt of fortifications, built in Belgium, built in 1859 to 1914 CE, running 95 km (59 miles).

See 'lines of Torres Vedras'.

Naval fire control tower. A German fire control tower for artillery, part of the Atlantic wall, an extant example can be found at L'Angle, on the island of Guernsey.

Naval turret. A turret taken from a ship and installed on a concrete bunker for land or coastal defence.

See 'turret'.

Nawabari. (J. *stretched rope*). The format of defences of a Japanese castle; the castle's plan.

'Nawabari' came in three forms:

(1)*'Renkaku'* style where the *'kuruwa'* are site along the ridges;

(2)*'Rinkaku'* contour style, where the baileys or *'kuruwa'* are placed concentrically, where the *'honmaru'* is surrounded by the *'ninomaru,* which is surrounded by the *'sannomaru'*;

(3)*'Teikaku'* style where the *'honmaru'* is sited closer to a mountain, sea or river (*'ushiro-kengo'*), the *'honmaru'* is adjacent to the castle's walls, and the additional baileys are sited around the *'honmaru'*.

See *'dôshinén'*, *'renketsukei'*, *'teikakukei'*.

Neck of bastion. The *'gorge'* or the interior side or entrance of a bastion.

See *'gorge'*.

Nejiro. (J. *root castle*). The main castle for military operations, as opposed to a *'shijou'* (branch castle).

Neribei. (J.). A variation of *'dobei'* constructed from used bricks and tiles mortared with clay and covered in plaster.

See *'dobei'*.

Nid-de-pie. (Fr. *crow's nest*). An entrenchment established on a breach by an attacking force to enable them to hold against counter attack, use dating from the 17th and 18th centuries CE.

Niederungsburgen. (G. *lowland castles*). As opposed to a *'höhenburg' or hill castle.*

Nijûhoriku. (J. *double trench*).

See *'hori'*.

Ninomaru. (J. *second circle*). The second defensive compound of a Japanese castle, which either surrounded or was adjacent to the *'honmaru'* or the main compound.

See 'dôshinén', *'maru'*,*'renketsukei'*, *'teikakukei'*.

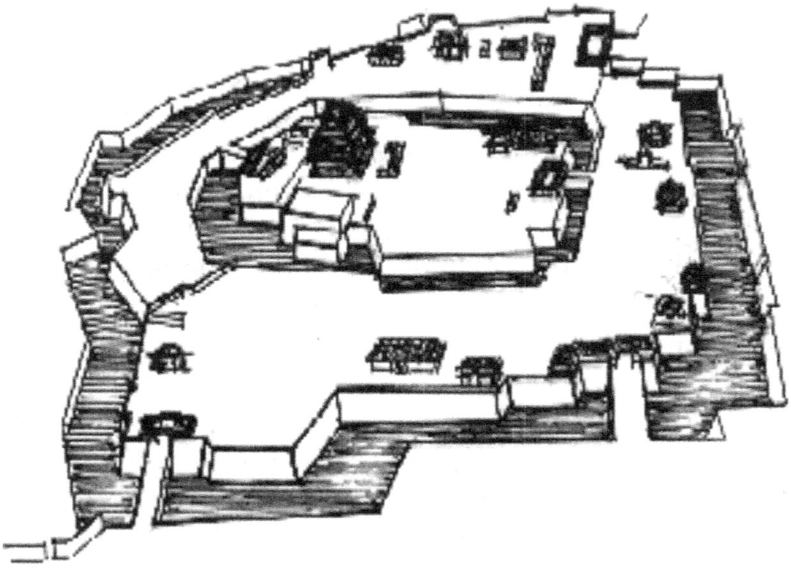

Nishi no maru. (J.). A compound (*'bailey'*) sited to the west of the main compound (*'honmaru'*).

See *'maru'*.

Nissen hut. Prefabbed steel half circle shaped hut used to house personal and equipment.

Nobori-ishigaki. (J.). Climbing stone walls. **See *'ishigaki'*.**

No man's land. The land between the opposing lines of trenches (WWI) through which troops had to pass to attack the opposing trenches. Also known as the *'zone of contention'*.

Nonozumi. (J.). Ishigaki where the stones are about the same size so that the stones line up across the face of the wall.

See *'nobori ishigaki'*, 'kikkozumi', *'kirikomihagi'*, *'nozurazumi'*, *'ranzumi'* 'tamaishizumi',*'tanizumi'*,*'uchikomihagi'*.

Nori. (J.). The slope of a stone wall or rampart of a Japanese castle.

Norcon pillbox. A British pillbox consisting of a non-reinforced concrete cylinder with multiple loopholes in the walls, later the design was modified to include a roof corrugated iron sheets which earth could be piled on. The pill box was cheap to produce and easy to build. There are twenty seven extant examples, with one at Moreton Ford, Dorset, England.

See 'British hardened field defences', 'FW3', 'pillbox'.

Noroshidai. (J.). A stone platform for a fire beacon for sending signals to other castles, usually on mountain summits, housing a dwelling for guards and somewhere to store dry firewood.

Northern frontier. ***See 'frontier nord'.***

North west front. ***See 'victoria lines'.***

Northumberland 'D' type pillbox. A British WWII pillbox unique to Northumberland in the shape of a 'D', similar to *'pillbox 24'*, an embrasure in each of the three forward facing walls, two embrasures on either side of the door in the long rear wall. There are seventeen examples extant including one near Eglingham, England.

See 'British hardened field defences', 'FW3', 'pillbox'.

Notched merlon. See *'Italian merlon', 'merlon'.*

Nozurazumi. (J.). Ishigaki where the stones are unworked or just simply split, this type of wall is not strong, not suitable for high walls and provides too many hand holds for enemy climbers.

See *'ishigaki', 'nobori ishigaki',* 'kikkozumi', *'kirikomihagi', 'nunozumi', 'ranzumi'* 'tamaishizumi',*'tanizumi','uchikomihagi'.*

Nri-durga. See *'durga', 'nara durg'.*

Nuclear shelters. Shelters provided protection from Nuclear weapon blast and radioactive fallout, some were just for the leaders but other were public (Scandinavian), providing long term accommodation.

Nuclear fallout shelters. China and Russia had a system of shelters for urban populations. Dutch and Swiss had a building code requiring space that was, or could be converted into shelters. Sweden had extensive underground fallout shelters. Britain had dispersed forces and submarines armed with missiles. The USA had no large scale civil defence scheme but a strategic nuclear missile force was housed in missile silos. An example is the Cheyenne Mountain Complex in Colorado, USA.

Numerus. A Roman fort used to house irregular troop units called *numeri* (such as the *numeri Brittonum*). An example is the '*numerus fort*' at Hesselbach in Germany.

Nunozuki. (J. *cloth construction*). Dry stone walling using rectangular shaped sideways in horizontal rows, leaving few gaps of climbers hand holds. Mainly used for foundations of castle gates, towers ('*yagura*').

See '*ishigaki*', '*nobori ishigaki*', 'kikkozumi', '*kirikomihagi*', '*nunozumi*', '*ranzumi*' 'tamaishizumi','*tanizumi*','*uchikomihagi*'.

Nu qiang. (Ch. *woman's wall*). The parapet of a wall, be it supplied with crenels and merlons or not. Also known as '*dié*' or '*pi ni*' (Ch. *Peek*).

Nurhag, Nuraghe, '*il nuraghe*' (double meaning of '*heap of stones*' and '*cavity*'). A megalithic structure of the Bronze Age Sardinia, consisting of conical stone structures, some 20 metres (60 feet) in height, some 7,000 Nuraghe have been found so far. Some Nuraghe can be multi-storied, provided with connecting passageways, stairs, and on whose sides are often open niches.

At Su Nuraxi di Barumini the Nuraghe consists of the central tower enclosed within four round subsidiary towers linked by a massive stone curtain wall. Surrounding the central tower are the remains of a second outer wall and a settlement of circular huts.

Three types of Nuraghe developed over time, including;

- Corridor, consisting of corridors and chambers, often with additional towers and features;
- Singular tower, conical in shape, also known as a '*Tholos*';
- Multiple round tower complexes with connecting walls.

See 'motilla' in Spain.

Nu tou. (Ch. *woman's head*). The merlon of a parapet. Also known as '*duo*' or '*dié*'. Chinese merlons were originally stepped, which were susceptible to damage to siege artillery and offered limited protection to those behind them. Later the merlons were rectangular in shape.

See '*nu qiang*'.

O.

Obiguruwa, obikuruwa. (J. *belt compound*). A narrow and long castle compound below the main compound of a mountain castle (*'yamajiro'*), usually built on top of an earthen embankment to form the outer edge of a moat. On flatland castles (*'hirajiro'*) the *'obiguruwa'* were narrow connecting areas flanking the moat.

See *'koshiguruwa'*, *'kuruwa'*.

Oblique embrasure. A type of embrasure that intersects a parapet at an acute angle.

Oblique fire. Incoming fire from an enemy that strikes the parapet at an acute angle, which is different to; *'direct'*, *'enfilade'*, *'plunging'*, *'reversed'* and *'ricochet'* fire.

See *'fire'*.

Obos. (Russian). Prefabricated wooden palisade.

See *'embrasure'*.

Observation post. A temporary or permanent position, used to gather information directing or providing support to operations, normally located in an elevated position.

See 'bartisan', 'cloche', 'escaraguaita', 'hochleitstand',' leitstand', 'luftschutzraum', 'fire control tower', 'observation tower', 'offener beobachter', 'tobruk'.

Obstacle. A passive or physical barrier to keep the enemy at bay, which can be natural like the type of terrain, lack or too much water, distance to travel, weather (season), density of plant growth, or man-made like ditches, moats, canals. Specialised man-made obstacles can be:
Anti-personal (*abatic, barbed wire'*, *'fraise'*, *'lilia'*, *'mine'*, *'pit falls'*, *'punji sticks'*);
Anti-landing craft (*'admiralty scaffolding'*, *'Atlantic Wall'*, *'belgian gate'*, *'hemmbalk'*, *'küstenvorfeldsperren'*, *'Rommel's asparagus'*);
Anti-tank (*'anti-tank obstacles'*, *'anti-tank mines'*, *'anti- tank ditches / trenches'*, *'anti-tank mines'*, *'dragon's teeth'*, *'Czech hedgehog'*, *'höckerhindernisse'*, *'tank traps'*);
Anti-glider (*'anti-glider trenches'*, *'Rommel's Asparagus'*. For an obstacle to work as designed they need to be maintained, defended and part of a wider defensive system, not merely moved out of the way or bypassed.

See 'abatis', 'barbed wire', 'beach obstacle', 'chevaux de fries', 'dragons teeth', 'entanglements', 'hairpin', 'hersillion', 'pariani posts'.

Octagon. An eight-sided fortification.

Offa's dyke. A linear earthwork consisting of ditch and rampart, roughly following the border between Wales and England, running about 193 km (120 miles) attributed to Offa, the King of Mercia (757-796 CE), but according to radiocarbon dating it is from 430 CE.

See 'Wat's dyke'.

Offener beobachter. (G. *open observer*). A small observation post built into the wall of German WWII bunkers. Also known as a '*tobruk*'.

See 'observation post'.

Offene bettung. (G. *open bed*). An open firing position, usually circular in shape with covered storage areas for ammunition, often made in concrete.

See '*bettung*'.

Ogi no kôbai. (J. *folding fan shape*). The '*ishigaki*' or stone wall of a Japanese castle that was concave in the transverse section, forming a parabolic shape. As the stone walls of a castle were made higher, the slope had to be changed to accommodate the increased stresses incurred. The concave shape also avoided the creation of dead ground at the base of the wall.

See '*ishigaki*'.

Ogival bastion. A bastion provided with a '*V*' shaped ogival point which was situated facing the most likely direction of attack, the ogival point added to the structural strength of the bastion, thus it was better able to resist battering by siege engine or artillery. An example is the 13th century CE Narbonne Gate, at the East entrance to the citadel of Carcassonne, France.

Ogival point. The sharp angled masonry projection of a tower which was used to eliminate dead ground at the base of the tower and add to its structural strength.

See 'en bec', '*ogival bastion*'.

Oilet, Oillet. The circular part of an arrow slit situated at the ends, the middle, the bottom or all of the above mentioned, which was used to give archers a greater field of fire. Also known as a '*loop*' or '*spy-hole*'.

See '*loop (2)*'.

Okajou. (J. *hill castle*).
(1) A castle built on a hillock rather than a mountain.
(2) A flatland mountain castle.

See '*hirayamajiro*', '*yamajiro*'.

Okkupationsburg. (G. *occupation castle*). A castle built on the border or in recently conquered territory to control the area.

Onager. A missile throwing engine, like a '*maginale*' but supplied with a short sling at the end of the throwing arm.

See '*mangonal*'.

Ondecagon. A fortification which has eleven sides.

Ongseong. (Korean). A '*chemise*' equivalent to a Chinese '*hu men qiang*'.

Oozutsu zama. (J.). Apertures in the walls of a castle for large cannon.

See '*sama*', '*zama*'.

Oote. (J.). The front of a castle.

See '*karamete*'.

Ootemon. (J.). The front entrance of a castle, usually in the form of '*yaguramon*'. Some castles may have had more of these types of entrances.

See '*karametemon*'.

Open back tower, Open tower, Open gorge. A tower, with an open back, so that if the enemy took the tower was open to fire from within. Used in the outer walls of Byzantine fortifications. Also known as a '*half tower*'.

See 'hollow bastion'.

Opere. (It.). Fort or work.

Opere da soldato. (It. *soldier's work*). The earthworks and other field works erected by the soldiers, like redoubts, counterworks and trenches. Also known as *'Arcbitettura soldatesca' (It. soldier's architecture).*

Open Flank. The portion of a flank covered by the orillion.

Oppidum. (L. *fortified town*). A large Iron Age town defended by ditch, rampart and palisade.

Opus Mixtum. (L.). The form of construction using alternating rows of brick and stone masonry, a feature of Byzantine walls, that is said to add to the strength of the wall.

Ordensburgen. (G. *order castle*). Castles built in Latvia and Estonia by the Teutonic knights to control territory during the Northern crusade in the 12th and 13th centuries CE. Built in red brick, rectangular in shape resembling cloisters and lacking towers.

See *'berg', 'bergen', 'burg'*.

Orechino. (It. *earing*). ***See 'orillon'.***

Ordu. A Turkish or Mongolian capital or palace which could be fortified, some consisted of concentric walls with towers.

Orgue. (Fr. *organ*). A timber obstacle made of timbers slid into grooves to close a passage of a gatehouse.

See porticullis'.

Oribei. (J.). A castle wall that is built with a zigzag pattern with loopholes in the flanks to reduce the amount of *'dead ground'* in front of the wall. Straight walls could have parapets built with the same zigzag pattern for the same reason.

See *'byoubuori'*.

Organisation Todt (OT). The OT was a German construction company during 1933 and 1945 CE (WWII) German army auxiliary force of construction troops (*'Bautruppen'*) initially consisting of German volunteers and registered building companies, later slave labour drawn from occupied territory. Used to build military infrastructure (i.e. the Atlantic wall, Flak towers, and concentration camps).

Orgue. (L.). A barrier made of thick planks of timber which slid in grooves closing a passage in a gatehouse.

See 'cha ban'.

Orillion, Orillon. An oblong projection at the shoulder of a bastion, projecting past the normal flank of a rampart and parapet; the curve of the orillion is convex to the ditch. Used to protect a flanking battery and defenders on the flank. Also known as an *'oblong bastion'*.

Ostrog (Russia *strogat*, '*to shave the wood*'). A small wooden palisaded fort made with sharpened logs, often built in remote areas or as part of the '*Zasechnaya_cherta*' (R. *Great Abatis line*).

See 'abatis'.

Ostwall. See '*Panther-Wotan Line*'.

Ôteguchi. (J.). Main entrance to a Japanese castle.

See '*ootemon*'.

Ôtemon. (J.). The main gate of a castle.

See '*mon*'.

Oubliette. (Fr. *oublier*, to forget). A dungeon with only one opening which was situated in its vault, usually covered by a trapdoor; a secret pit in the floor of a dungeon into which a prisoner could be thrown. An example can be found in the dungeon of Warwick Castle, Warwick, England.

Ougi no nawa. (J.). A plan of a Japanese castle in the shape of a folding plan, with the parts distributed on either side of the water supply for the castle. The main compound of a mountain castle ('*yamajiro*') would be located where the pivot for the fan would be. A wall or an earthen embankment was built along the edge of the water source to protect it.

Outer bailey. The bailey situated between the middle bailey or inner defences and the surrounding country; the defence which includes the main entrance, also known as the '*outer ward*'.

See '*barbican*', 'concentric castle'.

Outer curtain wall. The curtain wall surrounding the outer bailey.

See 'antemural', '*concentric castle*'.

Outer London defence ring. The defence ring consisting of a combination of pillboxes, anti-tank blocks, trenches, and minefields, with rivers and artificial ditches which was constructed around London in 1940 CE, part of WWII, to slow down the German advance on the capital.

See '*Lines of communication*', '*stop lines*'.

Outwork.

(1) The surrounding outer wall of a fortification.

(2) Defences constructed beyond the line of the main works, designed. to keep the enemy at a distance because of the effect of the enemy's projectile weapons (e.g. siege equipment or artillery); to break up the line of an assault; to cover the approaches to the other outworks; and when taken by an enemy force, leave them totally open to fire from the main works.

See 'advanced work', 'advanced line', 'antemural', 'antemuralia', 'arrow', 'barbican', 'counterguard', 'crownwork', 'detached work', 'forework', 'fléche', 'outwork', 'ravelin', 'fortalice', 'hornwork', 'horse shoe', 'moon', 'multivallate hillfort', 'paté', 'redoubt', 'russian sap', 'vamure', 'yue cheng', 'zwinger'.

Ouvrages. See '*works*'.

Oyajiro. (J.). The main castle.

See '*kojiro*'.

P.

Pa., A Maori stockade (New Zealand) containing stores for food in pits and huts of accommodation, and watch towers. The earliest Pa, Tiromoana at Te Awanga (Hawkes Bay, North Island) dates to 900 CE. A Pa consisted of a timber stockade with holes for thrusting spears through, with a terrace inside to enable those inside to throw missiles at the enemy outside.

The main idea of a Pa was to protect the tribes food reserves from other tribes, so damaging the Pa was not the idea but winning also had a spiritual aspect.

Pagan, Blaise François. (1604 – 1665 CE). A French military engineer serving under Louis XIII, published *'Les Fortification'* which was an influential text which influenced the work of '*Vauban*'.

See '*Vauban*'.

Palanka. (T. from Greek *'phálanks' for log)*. A Turkish fortification used in Southern Europe against Austria and Hungary, of various shapes and sizes, which could be built quickly, and expanded swiftly. The timber could later be replaced by stone. Palankas could be built adjacent to a town or an extension of a main fortress. *Also known as 'parkan' or 'palanga'.*

Palintonon (Gr.). A Greek heavy catapult for throwing stones. Also known as a *'lithobolos'*.

Palisade. (L. *palûs*, a stake).

(1) A defensive fence or wall of wooden stakes arranged vertically or obliquely in a row.

See *'valli'*.

(2) To surround or defend with a palisade.

Pamart armoured cupola. ***See 'cloche pamart'.***

Pan. The part of a work forming one side of a salient angle.

Panchalika. (Sanskrit).

(1) A wooden platform floated in a moat provided with spikes used as an obstacle to enemy attackers.

(2) A wooden plank studded with nails used as an obstacle in a ditch to impede enemy attacks.

See *'hersillon'*.

Pan coupé. When a salient angle is replaced by the straight line of a parapet, a pan coupé is formed.

Panka-durga. (Sanskrit). An Indian fort surrounded by marsh and quicksand.

See 'durga'.

Panther-Wotan Line. (G.). German defensive line built in 1943 CE on the Eastern Front with Russia. a combination of natural barriers with trenches, bunkers, anti-tank obstacles, and fortified positions. Partially built and poorly supplied, manned and was overrun in 1944 CE. The line ran between Lake Peipus and Narva on the Baltic Sea, then along the Dnieper to the Black Sea. Also known as '*Ostwall*' (G. *East wall*).

Panzerabwehrgraben. (G. *anti-tank trench*).

Panzerdeckungsgräben. (G.). Tank cover ditch.

See '*trench*'.

Panzerdeckungslöchern. (G. *armour cover holes*). Armoured protection hole or trench.

Panzerfallen. (G.). Anti-tank ditch.

See '*trench*'.

Panzerfort. (G. Tank fort).

Panzerkampf wagenstand. (G.). A dug out position for a tank in the '*turret up*' position, with sides re-enforced by timber poles holding '*fascines*' in place.

See '*tank defilade*'.

Panzerlafette. (G. *tank mount*). German armoured mounting design in the late 19th century consisted of a small cylindrical domed armoured turret on wheels, armed with a quick firing gun.

Panzernest. (G. *tank nest*). German WWII small armoured mobile prefabricated pillbox made of cast steel, used in Russia and Italy, armed with a machine gun.

See 'pillbox'.

Panzersperre aus aumstämmen. (G. *tank barrier made of tree trunks*).

See 'abatic', 'astverhau'.

Panzerstutzpunkt (PzStP). (G. *tank base*). German WWII armoured stronghold which included concrete bunkers ('*panzerwerk*') with artillery placed in turrets (*panzerturm*).

See 'hedgehog defence'.

Panzerturm (20P7). (G. *armoured turret*). An armoured bell of the Siegfried Line made from chrome molybdenum steel (250 – 295 mm thick (98 -116 inches), weighing 56 tons (50.8 tonnes), 2.25 metres (7 foot, 5 inches high), used as machine gun firing positions, which would have been partially buried in the ground.

Panzerwerke. (G. *tank work*). A German WWII bunker with artillery placed in turrets (*panzerturm*).

Parados. (Fr. *parare*, to prepare; L. *dorsum*, back). An earthwork parapet situated behind an infantry or gun position, protecting the rear from attack. Usually shored up with timber planks and sandbags.
Parallel. A siege trench dug parallel to the front of a fortress.

See *'trench'*.

Parapet. (It. *parapetto*).
(1) The top of a wall of either a fortification or fieldwork, either plain or battlemented. Used to provide protection to the defenders behind the wall.

See *'battlement', 'crenel', 'embrasure', 'merlon', 'reveal'*.

(2) A breastwork or wall used to protect the defenders on the ramparts of a fortification, either plain or provided with embrasures.
Parapet walk. See *'allure'*.
Paraportia. Gates in a Byzantine field fortifications. See the *'Strategikon'* for more details.
Parcham, Parkam. (L. *parricus*, park). Was used instead of a *'zwinger'* in the territories of the Teutonic Order.

See *'zwinger'*.

Pareclat. A wall raised on a rampart or in the middle of a bastion or a trench, of either earth, gabions or sandbags, used to absorb shrapnel.

See 'parados', 'traverse'.

Pariani posts. An Italian casement developed in 1938 CE, consisting of a single concrete block with metal armour, designed to withstand small to medium calibre weapons, armed with one or two machine guns, sometimes anti-tank guns. Named after General Alberto Pariani. Also known as a *'7000-type work'*.

See 'alpine wall', 'casemate'.

Parigha-durga. (Sanskrit). An Indian fort surrounded by earthen, stone and brick walls.

See 'durga'.

Parighas. (Sanskrit). Iron spikes on the fronts of gates of forts to prevent attack by elephants.
Parikha. (Sanskrit). A ditch or moat around a fort. The spoil from digging the ditch is used to raise the rampart or *'vapra'*.
Parmerston Forts. A number of forts were built around the coasts of the United Kingdom and Ireland following the recommendations of the 1860 CE Royal Commission on the Defence of the United Kingdom in response to the threat of French invasion. The varied works were also known as *'Palmerston's Follies'*, as the costly ornamental building was of

no intrinsic value due the fact the threat from France had passed and the structures themselves, let alone the guns, became obsolete technology, a waste of public monies.

See '*martello towers*'.

Parol. A wooden frame with sharp stakes projection horizontally from it, which was used to deter besiegers from scaling the parapets.

Pas de souris. (Fr. *pas de souris*, no mouse).

(1) A removable wooden stairway from the '*covered way*' or '*place of arms*' to the ditch of a permanent fortification, the removal of the stairway would be an obstacle to an attacking enemy.

(2) A set of stairs providing access from the ditch to the terreplein of the outworks.

Pashtib. (Indian). A raised platform of sandbags used to fill a defensive ditch.

Passage. An opening in the parapet of the covered way.

Paté. An outwork in the shape of a horseshoe, used to cover the gate of a fortification.

See '*demilune*', '*halfmoon*', '*horseshoe*', '*moon*'.

Patio de armas. (Sp. *parade ground*). The courtyard or bailey of a Spanish castle.

See '*bailey*'.

Pechnase. (G. *pinch nose*). **See '*machicoulis*'.**

Pele, Pele tower, Pile tower. (L. *palus*, stake). A castle almost unique to the Scottish/English border region, usually consisting of just a small tower or fortified dwelling, suitable for defence against raiding parties rather than any type of real siege. Dating from 1100-1700 CE, they are also found in Ireland, used both as a refuge and a stronghold. Some were provided with a surrounding or adjoining bailey which was called a '*barbkin*' or '*barmkin*', which was used for the protection of the tenants and their livestock during raids.

Penthouse. See *'testudo'*.
Penstock. A wooden dam, the water of which was used by the defenders of a fortification to destroy enemy positions in the ditch by opening the sluice of the dam and flooding the ditch.
Pentagonal pillbox. A British WWII pillbox supposedly for the defence of airfields, regular pentagon with embrasures in each face, door and embrasure in rear, variation in size. An example can be found in Alton Barnes in Wiltshire, England.
See 'British hardened field defences', 'FW3', 'pillbox'.
Pentagonal sconce. An earthwork fort of the 17th century CE, based on the bastioned trace plan, protected by; pitfalls and a palisade beyond the ditch surrounding the sconce, and horizontal storm poles on the top of the interior slope of the ramparts, as well as a wooden revetted breastwork which extended along the top of the ramparts.
See *'bastion trace'*, *'sconce'*.
Pentis, Pentice, Pentise. (O.Fr. *apentiz*). A covered walk-way along a wall, or just a shed or lean-to added to a building.
Perekop wall. (Slavic. *dug out*). A ditch and rampart separating the Crimean from the mainland from Sivash Bay (Azov Sea) to Karkinitsky Bay (Black Sea), about 10 km (6 miles) long, with the remains of a fortress at each end, also known as the '*Turkish wall*' since the Turks built fortification there in the 15th and 16th centuries CE.
See 'linear defences'.
Peribolos. (G.). Enclosure, court or ward.
Perimeter wall. See *'bailey wall'*, *'curtain wall'*, *'enceinte'*.

Permanent fortifications. A well built, durable defences of a place, intended to last a long time, used to defend a cities gates or some other fixed strategic location.
Perpendicular. The line at the centre of a side of a polygonal fortification, used to establish the positions of the *'lines of defence'*.
Perpendicular fortification. A fortification introduced by Montalembert in 1776 CE, consisting of four different lines of defence which were; firstly, a tenaille trace, followed by two lines of counter guards, and finally a two storied circular multi-gun tower. Thus, four consecutive breaches were needed to be made to subdue this type of fortification.
Perrier. See *'trebuchet'*.
Petard. (Fr. *'péter'*, to break wind). A type of small charge made of a metal or wooden box filled with gunpowder attached to a gate, door or wall, used to make a breach, which were put in place by a soldier while under the defenders fire. If the charge went off prematurely and injured the soldier who was handling the petard, it was said that they were *'hoisted by their own petard'*.
Petrobolos. (Gr.). A stone throwing catapult.
See *'palintonon'*.
Pettah. A Sri Lankian town attached or adjacent to a fortress, sometimes could also be fortified.
Petraria, Petraries, Petrary. Stone tossing siege engine used by Romans and in the Medieval period.
Pfahlsperre. (G. *pile barrier*). A barrier made from logs or concrete posts.
See 'abatic', 'anti-tank posts', 'astverhau'.
Pfostenschlitzmauer. (Gk. *post slot wall*). A defensive wall made of vertical wooden posts forming a lattice which was filled with earth and rubble. Also known as a timber framed wall, murus gallicus or gallic wall, used to protect Iron age hill forts in Europe.
Phroura (Gk. *guard*). A fortified citadel.
Phrourion. (Gk. *fortress*). A fortified collection of buildings used to house a military garrison.
Piatta forma. (It. *platform*). A type of flat bastion built on the curtain between two bastions to increase the flanking ability.
Piazza bassa. A casemate set in a flank of a work from which artillery was used to enfilade the curtain and to cover the ditch.
Picric acid (Gk. *pikros*, bitter). An organic compound also known as 2,4,6-trinitrophenol. Used in dyes and explosives.

Picket, Picquet, Piquet. (Fr. *piquer*, prick).
(1) A pointed stake.
(2) Set place(s) with stakes.
(3) A one man *'rifle pit'*, *'skirmish hole'*, *'fox hole'*.

Picket fence. A fence made from pickets or stakes.

See *'palisade'*, *'picket'*.

Picket-Hamilton pillbox. A British WWII pill box used to defend airfields in 1940 CE, consisting of a retractable turret, armed with light machine guns, accessible via a metal hatch in the roof.

See 'FW3'.

Picquets. Used by the English in India, set up in the outer ring of camps, manned by a platoon. Protected by a stone breast work (or *'sangar'*) plus a light trip wire.

See *'picket'*.

Pictish tower. A tall round tower dating from the early Iron Age which was attributed to the Picts. Chiefly found in the Shetlands, Orkneys and Northern Scotland.

See *'broch'*, *'dun'*.

Pieced rampart. A rampart which was modified by replacing the crenels with embrasures so that cannons could be fired through the parapet.

See *'embrasure'*, *'rampart'*.

Piéce maitress. (Fr. *master piece*). The French phrase for the architecturally dominant feature of a castle, which in most castles was the keep or donjon, or can be applied to the great gatehouse which took the position of the keep.

Pila muralia. (L.). A Roman beam sharply pointed at both ends with a reduced section in the centre acting as handle and aid in tying other pila together, used in fortification.

See 'sudis'.

Pile barrage. A barrier in a body of water used to deny access to water craft, consisting of wooden stakes (or other material) driven into the bed with the tops reaching to or above the water line, could be used in conjunction with floating booms (i.e. the Shoeburyness Boom, Thames Estuary, United Kingdom).

Pill boxes. Small forts with embrasures for the defence of stop lines, canals and air fields.

In the UK there were about 8000 pill boxes built for WWII, in various types, however there were poor records of locations by the Public Records Office.

Various designs included: Hexagonal Bren gun emplacement; Rectangular emplacements for land and air defence; Square and round form; Anti-tank 2 pounder emplacement; Sea wall pattern. Some were camouflaged like those among the ruins of Pevensey Castle in the English county of East Sussex, or a beach side kiosk.

See 'blockhaus, 'Casemate type G', 'Dover quad pillbox', 'Eared pillbox', 'FW3', 'Moir pillbox', 'Mushroom pillbox', 'Norcon pillbox', 'panzernest', Picket-Hamilton pillbox, 'Ruck pillbox'.

Pillbox 22. The most common British WWII pillbox (with 1,346 extant), a regular hexagon in plan, with embrasures in five of the sides and a door in the 6th side, the embrasures were suitable for light machine guns and rifles.

See 'British hardened field defences', 'FW3', 'pillbox'.

Pillbox 23. A British WWII rectangular pillbox made up of two squares, the front has three embrasures and no roof, the embrasures were suitable for light machine guns and rifles, the open section could also be used for light anti-aircraft defence, the rear section was roofed. Type 23 and its variants were rare with eighty seven as being recorded as extant.

See 'British hardened field defences', 'FW3', 'pillbox'.

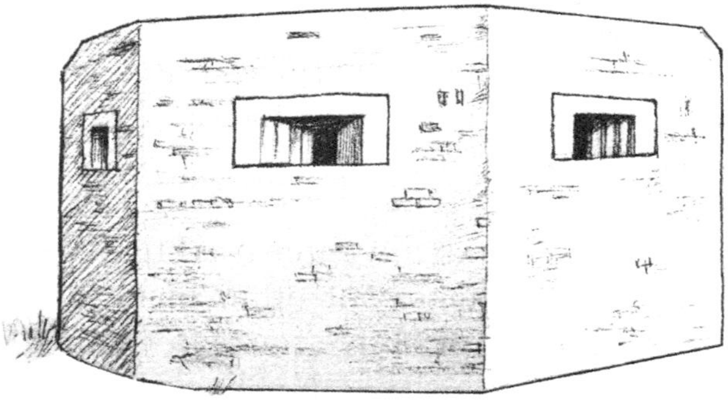

Pill box type 24, with stepped embrasures.

Pillbox 24. Similar to *'pillbox 22'* but is an irregular hexagon with the rear wall the longest, with a *'Y'* shaped wall to reduce the effects of ricochets. Type 24 was the most common with 1,724 recorded as being extant.

See 'British hardened field defences', 'FW3', 'pillbox'.

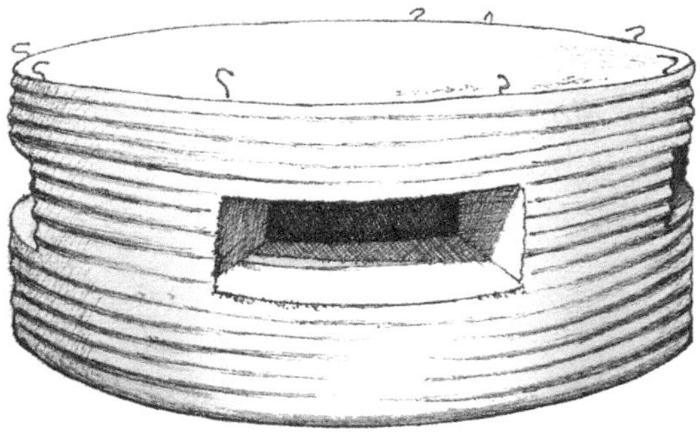

Pill box type 25, with hooks for attaching camouflage netting.

Pillbox 25. The only circular pillbox design with three embrasures of rifles and light machine guns, and a small entrance on the rear. Type 24 was rare with thirty recorded as being extant.

See 'British hardened field defences', 'FW3', 'pillbox'.

Pillbox 26. A square British WWII pillbox with a door on one side and three embrasures for rifles and light machine guns, occasionally there was an embrasure next to the door entrance. Type 26 also had a prefabricated version made by The Stent Precast Concrete Ltd. Type 26 was uncommon with one hundred and forty recorded as being extant.

See 'British hardened field defences', 'FW3', 'pillbox'.

Pillbox 27. The largest of the British WWII pillboxes, octagonal in plan, seven embrasures for rifles and light machine guns, with an open well in the centre for a light anti-aircraft position, and a protected entrance. Type 27 was uncommon with eighty four recorded as being extant.

See 'British hardened field defences', 'FW3', 'pillbox'.

Pillbox 28. The largest of the British WWII pillboxes with an anti-tank design, square in plan, and the large front embrasure was stepped and the gun shield would mostly fill the embrasure. The other embrasures could be defended by rifles or light machine guns. Type 28 and its variants were common with three hundred and six recorded as being extant.

See 'British hardened field defences', 'FW3', 'pillbox'.

Pi lian. (Ch. *leather curtain*). A temporary crenel shutter made from leather.

See '*xuan hu*', '*xuan lian*'.

Pioneers. (Fr. *pionnier*, foot soldier). Soldiers that were provided with a range of tools (axes, billhooks, mattocks, picks, spades, saws) to clear obstacles in the way of an advancing army, or to build field works.

See 'engineers', 'miners', 'sappers'.

Ping tou qiang (Ch. *flat top wall*). A rare type of parapet that was not provided with crenels but with loop holes.

The town wall of Sanhe, Anhui, China.

Pinna, Pinnae. (L.). A merlon on top of a breastwork of a Roman defence work such as at Alesia.

Pile barrage. A barrier in a body of water used to deny access to water craft, consisting of wooden stakes (or other material) driven into the bed with the tops reaching to or above the water line, could be used in conjunction with floating booms (i.e. the Shoeburyness Boom, Thames Estuary, United Kingdom).

See 'coastal fortification'.

Pile tower. See *'pele'.*

Pisa. A revetment made using sun dried bricks.

See 'revetment'.

Pit falls. Earthen pits dug into the ground outside a fort, provided with sharpened wooden stakes or spears at their bottom. Examples: Rough castle on the Antonine Wall, Central region, and the Queen's Sconce at Newark, April 1646 CE, outside the work.

See *'lilia'.*

Place. A word used in military architecture when referring to a fortification or a fortified town.

Place des armes, Place of arms. An enlargement of the covered way where troops could assemble for sorties, and were used as command posts; if the work was salient it was situated at the outer point of the covered way, while if the work was re-entrant it was situated at the inner point.

Plane of sight. A line sighted by an engineer that represented the converging enemy's fire into the interior of a fortification.

Platform. A foundation supporting an artillery piece, usually built of timber, which stopped it digging itself into the surface of the '*terreplain*'.

See 'terreplain'.

Plinth. The original purpose of the plinth was to increase the effect of the machicolations, and was not so much to counter the endeavours of sappers as to deflect projectiles thrown down from the walls above. The upper section of the plinth was concave and the lower oblique; the effect of this shape deflects projectiles thrown from above in a fan-shaped trajectory.

See *'batter', 'spur', 'talus'.*

Plongee. The downward slope of the superior slope of a '*parapet*', or the downward slope of the floor of an '*embrasure*'.

See 'embrasure'.

Plunging fire. Artillery fire in a high arc from mortar that enables a shot to land behind a place, which was more effective with explosive projectiles.

See *'fire'.*

Plutei, Pluteus. (L).

(1) A mantlet or mobile shed formed of wicker, used to protect besiegers attacking a fortification.

(2) A breastwork or parapet.

Pointed arch machicolation. A machicolation dating from the 14th to the 15th century CE, which was supported on buttresses which extended down to the foundations. The type of machicolation had two distinct advantages. that larger projectiles could be cast down through them onto rams and sappers below; and should the sappers succeed in making a breach in the wall, the buttresses and arches would localize the effects of the damage.

See *'machicolation'.*

Point blank fire. Where the weapon's fire is directly at the object being attacked.

See *'fire'.*

Pointed bastions. See '*en bec*', 'ogival bastion'.

Pointed merlon. A merlon typical of the Mudéjar military architecture, having a pointed apex.

See *'merlon'*, *'Moorish merlon'*, *'Mudéjar merlon'*.

Pol. (Sanscrit). A gate or an entrance to a closed area.

Poliorcetics. (L.). The art of siege craft.

Poliorketès. (Gr.). Besieger.

Poliorkia. (Gr.). Siege.

Polverista. (It. *gunpowder*). A gunpowder magazine.

Polygonal fortification. A fortification designed by Montalembert in 1777 CE which consisted of faces with salient angles of little depth, and powerful *'caponiers'* armed with twenty seven cannons in casements and a large number of infantry loopholes.

Polygonal keep. A keep which was built on a polygonal plan and was designed to overcome the problems associated with square and rectangular keeps, whose corners were vulnerable to battering, and their cubic shape created dead ground by their reduced field of fire, and they were also more susceptible to attack by undermining.

See *'keep'*, *'transitional keep'*.

Pomerium, Pomoerium. A street like open space which runs parallel to the inside base of the fortified wall of a city.

Pontoniering. Construction of temporary military bridges or ferries by *'engineers'* and *'sappers'* using pontoons.

Port. (L. *porta*, gate). A slit or aperture in the walls of a fortification through which guns could be fired at an enemy force, used to provide flanking and covering fire.

See *'gun loop'*, *'gun port* loop'.

Porta decumana. The rear gate of a Roman fortification, situated in the wall opposite the main gate (*'porta preatoria'*) usually situated in one of the shorter walls.

Porta praetoria. (L.). The main gate of a Roman fortification, normally situated in one of the shorter walls opposite the *porta decumana* or rear gate.

Porta principalis dextra. (L.). The main gate on the right hand side of a Roman fort connected by the *'via principalis'* to the *'porta principalis sinistral'*.

Porta principalis sinistral. (L.). The main gate on the left hand side of a Roman fort. connected by the *'via principalis'* to the *'porta principalis dextra'*.

Portcullis. (L. *porta*, door and *coleice*, sliding). A heavy grating of iron or iron reinforced wooden bars, suspended on chains which were worked by winches; used to cover an entrance to a castle. The portcullis could be dropped quickly down vertical grooves in the entryway called *'coulisses'*, or at any other important entrance in the event of attack. Some castles had the machinery for raising and lowering the portcullis, usually a windlass, was accommodated above the entry, in a chamber situated between the flanker towers and was only accessible through one of them.

This chamber was often put to domestic use despite the obstruction of the windlass and often the machinery for moving the drawbridge as well. Some gateways were provided with a succession of portcullises, at some distance apart.

See *'gatehouse', 'successive portcullis'.*

Posidonia. (Gr.). Ditch. (i.e. Paestum).

Positional warfare. A military doctrine with the emphasis on control of terrain and territory by fortifications and obstacles, thus denying the ground to the enemy.

See 'Atlantic wall', 'Great wall of China', 'Hadrian's wall', 'trench warfare'.

Postern, Postern gate. Old French posterne, alteration of posterle, from Late Latin posterula (*'back door'*), from Latin posterus (*'later'*). A small secondary entrance, sometimes concealed, and usually at the rear of a castle. Used as a sally port for sorties, and as a route of escape.

See *'sally port'.*

Postern gate (the Theodosian Land walls of Constantinople). Hexagonal tower south of the Yedikule complex / Golden gate. The postern gate is hidden in the corner between tower and curtain wall, and is provided internally with putlog holes to bar the postern.

Postern bridge. A bridge crossing a moat or ditch from the postern gate.

See *'postern'*.

Postern tower. A tower which was used to defend as well as sometimes incorporating the postern gate.

See *'postern'*.

Poterne. (Fr. *postern*). An underground masonry passage connecting the gorge to other subterranean parts of Dutch 19th century CE fortifications, such as barracks, cupolas and caponiers.

Poudrière. (Fr. *powder magazine*).

Powder magazine. A specific building for the safe storage of gunpowder (i.e. Berwick–upon-Tweed, England, UK).

Pradhavitika. (Sanskrit). A place in a fort where a garrison could be hidden.

Pragandi. (Sanskrit). A temporary construction on the parapet with holes in the bottom through which missiles could be dropped on attackers.

See ***'hoarding'.***

Prakara. (Sanskrit). A type of parapet supplied with embrasures for loosing arrows, made from stone or brick.

Prantara-durga. (Sanskrit). A fort sited on a flat hill summit.

Pré carré. (Fr. *duelling field*). Developed by Vauban in 1678 CE for King Louis XIV of France, along the northeastern border of France, consisting of two defence lines of connected forts for mutual defence, from the sea to the Ardennes, which included a programme of reducing the number of unneeded forts, saving money and spending the money on the required forts. Also known as the '*Frontière de fer*'.

See '***Vauban***'.

Precinct. The space enclosed by fortified walls.

See ***'bailey', 'enceinte'.***

Presidio. (L. *praesidium*, garrison). The Spanish term for a place of defence such as; a fort, garrison or garrison town, or just a guard house.

Priest cap. A fortification; so named because of its similarity to the shape of a priest's cap, also known as a '*mitre*' or a '*swallow tail*'.

Principal magazine. The main powder magazine of a fort sited as far as possible from the parts of the fort exposed to attack.

Principia. (L.). The headquarters of a Roman fortress or auxiliary fort which was situated at the junction of the '*via praetoria*' and '*via principia*'.

Profile. The outline of a vertical section of work or cross section, containing '*trenches*', '*rampart*' and '*ditch*'.

Promenthouse fortified line. One of the fortified lines in Switzerland, 10km (about 6 miles) long, between Bassins and Prangins in the Canton of Vaud, consisting of lines of '*dragon teeth*' and twelve fortresses (i.e. Villa Rose). Also known as the '*Toblerone line*' because the concrete blocks look like the brand of chocolate. Construction took place in 1939 CE.

See '***dragon's teeth***'.

Promontory fort. See *'cliff fort'.*

Propugnaclum. ***See 'hoarding'.***

Proteichisma. (Gr.). Ditches combined with outworks.

Protoli. (Sanskrit).

(1) A small gate adjoining the main gate.

See *'postern'*, *'wicket'*.

(2) A two story construction between *'attalakas'* wall towers.

Prussian system. Based on the polygonal trace, using casemated keeps, caponiers, Carnot's walls, and concealed mortar batteries. An example is Fort Friedrichsau, Germany, 1850's CE.

Pulley drawbridge. A drawbridge raised and lowered by ropes or chains via a pulley system.

See *'drawbridge'*.

Pulverturm. (Gk. *powder tower*). A tower for the storage of gunpowder. Also known as a *'pulverhouse'* (Gk. powder house).

Punic ditch. A defensive *'V'* shaped ditch with one side much steeper than the other; the ditches surrounding a fortlet had an outer face at 27 degrees and the inner face at 69 degrees. The rampart behind the ditches only survives to 0.5 metres (1.6 ft) at its highest point.

Punji sticks. Anti-personal field device consisting of sharpened bamboo sticks in a pit which would piece combat boots.

Put-log. A log used to block a door or gate way, put into the *'putlog holes'* in the stone work on either side of the door way.

Put-log holes. Holes in the stonework on either side of a door or gate to secure a put log to brace the door or gate if attacked. One hole is square to take the end of the log, while the opposite hole is *'L'* shaped so the end of the log can be slid in and down into place.

Pyramid.

(1) A Dutch concrete observation bunker built before WWII (*'Groepsshiuiplaats'*, Group shelter) with sloping roof like a pyramid, a single entrance with an armoured door, a staircase leading to a corridor defended by a rifle hole (*'schietgat'*), leading to the standby room (*'afwachtingsruimte'*) which had a ventilation system, telephone, and periscope.

(2) ***See 'Anti-tank pimples'. 'dragon's teeth'.***

Q.

Qalat. (Ar.). Mudbrick fort or defensive compound common throughout southwest Asia.

Qalca. A small tower in the mountains of the Iberian Peninsula with large storage facilities. Also known as a '*burj*' or a '*qubba*'.

Qasabah, Qasr. (Ar.). A governor's fortified residence and garrison.

See 'alcazabar', '*ksar*'.

Qian jin zha. (Ch. *thousand jin floodgate*). **See '*cha ban*'.**

Qi Cheng. (Ch. *banner city*). ***See 'man zhou cheng'.***

Qi cheng pu. (Ch. *wall riding shack*). A '*cheng pu*' for narrow walls with not enough room for a normal '*cheng pu*', more a '*bridge tower*' or '*chemin de ronde*'. The arched gateway allowed defending troops to pass under the tower. Living quarters were provided on the first floor and above. Also known as '*qu cheng zou ma pu*' (*wall riding, horse walking shack*).

See '*cheng*', '*cheng pu*'.

Quadrangular castle. A castle dating from the 14th century CE, which consisted of four wings surrounding a central courtyard. The outer face was reasonably defensible, the corner towers were substantial, while the interior faces were almost totally domestic in character. The quadrangular castle combined display, comfort and security and was the dominant form of English fortified dwelling from the middle 14th century CE until the time they merged with the great undefended quadrangular houses of Queen Elizabeth 1st's reign.

See 'courtyard castle', 'keepless castle'.

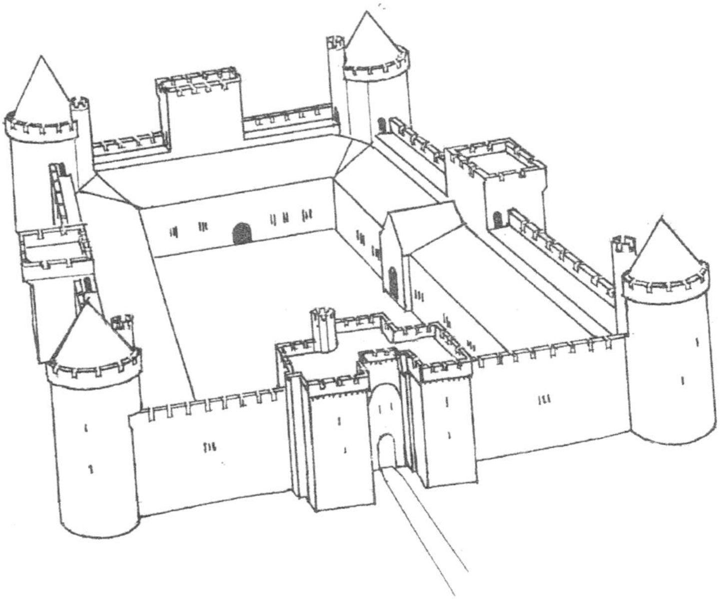

See Appendix 1. What is a Castle?

Quadrant castle. A castle on a site formed by constructing a line of defences between principal natural obstacles and secondary obstacles, such as a tributary or lateral ravine.

Quadriburgia.

(1) A Frankish castle in the Levant, consisting of two square enclosures, one inside the other, thus a concentric without a keep.

See 'concentric castle'.

(2) Roman rectangular forts for auxiliary units defending the '*Strata Diocletiana*' (the Road of Diocletian, 284 to 305 CE) a fortified road along the eastern desert border of the Empire from the southern bank of the Euphrates river passing Palmyra and Damascus to northeast Arabia.

See 'strata diocletiana'.

Quadrilateral castle. See '*quadrangular castle*'.

Quatrefoil keep. A shell keep which was usually situated on an artificial motte and consisted of four overlapping circular bastions. The blind angles at the joints of the bastions were covered by bartizans supported on corbels, providing flanking fire rather than vertical defence. The design was derived from a type of French keep, which provided an all-round field of fire, and a similar design was later adopted for the artillery forts of the Tudors. Also known as a clover-leaf keep or a four-leafed flower keep (i.e. Deal castle, Kent, UK).

See 'device forts'.

Quaker guns. Fake cannons made from black painted logs, used to produce a false impression of a fortification's fire power, used in the 18th and 19th century CE.

Quang Ngai. ***See 'long wall of quang Ngai'.***

Quatriburgia. Roman (square) frontier fort from the Euphrates River, through Syria, Transjordan and down to the Red Sea. An example is at Yotvata (Israel) guarding the road from Gaza to Aila.

Qubba. See '*qalca*'.

Que tai. (Ch. *magpie platform*). A raised platform or secondary wall with its own battlement ('*Ping tou qiang*') usually built on top of a main wall or near a moat. Previously used to withstand stone throwing artillery, later became obsolete with the advent of stronger walls and better battlements.

Queue d'aronde. (Fr. *dovetail*). An advanced work in the shape of a tenaille with two wings.

See 'priest cap'.

Quincuncial disposed arrow slits. Arrows slits placed in a pattern of five to avoid weakening the masonry and to increase the field of fire.

See '*staggered arrow loops*'.

R.

Radar installations. Installation used to detect and warn of incoming enemy aircraft with radar first used in WWII.

Railway guard post. A post used to protect valuable points of a rail system, like bridges, depots and switching points, using stockades, block houses, redoubts and entrenchments.

See 'rice blockhouse'.

Rainures. (Fr. *grooves*). The vertical slots in the front of a gatehouse allow for the movement of the gaffs (timbers holding the chains) of the bascule bridge of a castle.

See 'bascule bridge', 'drawbridge', 'gaffs'.

Raking shot. Cannon shot which sweeps across a target.

See 'fire'.

Ram (Germanic). A tool used to make a breach in a gate or wall which could be just a tree trunk man handled by soldiers, in the middle ages rams were strengthened with metal bands and heads. The ram and those using it could be protected by a mobile roof (mantelet, penthouse, siege tower), where the ram could be suspended above on chains or ropes and easily swung to and fro at the target.

Ram (defences). Device to attack a ram could include:

(1) Hooks on ropes or a noose used by defenders to catch the ram and be used to overturn the whole engine.

(2) A beam swung on a pulley system to strike the ram.

Ramp.

(1) An inclined plane joining two levels of an earthwork or wall.

(2) To construct a ramp.

Rampart. (O.Fr. *rempar*, to defend). An embankment of earth which was used for the purpose of defence, excavated from the ditch, and either raised on the inside or outside of the ditch. A number of different types of ramparts have been employed, generally they surrounded a fortification and were usually topped by a parapet.

See *'agger', 'dump or glacis rampart'*, 'revetted earthen rampart', *'turf rampart', 'timber laced rampart'*.

Rampart plane. The part of the rampart that represents the converging fire along the length of the ramp and commands the heights in front of the work.

See 'rampart'.

Rampart slope. The side of the ramp between with the '*banquette*' or the '*terreplein*' and the rear of the work.

See 'rampart'.

Rams horns. A low work constructed in the ditch in the shape of circular arc, invented by M. Belidor, which were used instead of tenailles.

Randhausburgen (G. *surrounding-house-castle*). A German castle with the parts of the castle built around a courtyard (i.e. Eltz, Germany, 12th to the 16th century CE).

Ransekizumi. (J.). The piling of unhewn rock into a dry stone castle wall. Also known as '*ranzumi*'.

See '*gobouzumo'*, 'ishigaki', 'nobori ishigaki', 'kikkozumi', 'kirikomihagi', 'nunozumi', 'ranzumi' 'tamaishizumi','tanizumi','uchikomihagi'.

Ranzumi. (J.). '*Ishigaki*' where the stones are of various sizes and there is no actual pattern across the face of the wall.

See '*nobori ishigaki*', 'kikkozumi', '*kirikomihagi*', '*nozurazumi*', '*nunozumi*', 'tamaishizumi','*tanizumi*','*uchikomihagi*'.

Rasant. Sweeping or grazing fire. ***See 'fire'.***

Rath. An Irish earthwork fortification dating from the Iron Age.

Rathya. (Sanskrit). Chariot roads built within a fort for swift movement about the fort.

Ravelin. (It. *ravellino*). A detached outwork developed from the demilune, consisting of a triangular work with two embankments raised before the counterscarp, the work itself was isolated in the ditch of moat. One purpose of the ravelin was to shield the entrance to a fortification from direct bombardment. The ravelin was accessible either by a drawbridge if it formed a part of the road system of a fortification, or if only part of the defence works access was by a tunnel or timber bridge from the inner works. The earliest ravelin was built by Saranello between Leghorn (Livorno) and Genoa in 1377 CE.

See '*demilune*'.

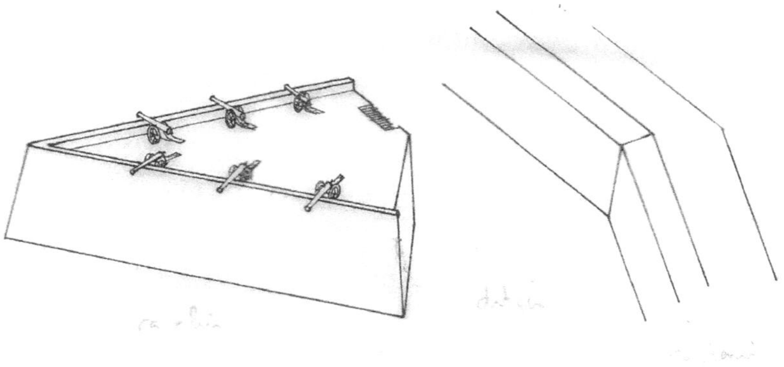

Rear bonnet. A work at the rear of a fortification facing the fortified square, opposite to the direction of enemy attack.

See 'bonnet'.

Rectangular buttress turret. The rectangular shaped flanking turret which only projected a small distance from the four corners of a Norman rectangular or square keep (i.e. the Tower of London).

See '*corner buttress turret*'.

Redan, Reden. (*Re* + L. *den*, tooth).

(1) A field work consisting of two faces and an open gorge.

(2) A triangular work situated forward of the main fortification, consisting of two faces and an open gorge, like but larger than a '*flèche*'. Used to fortify walls when the necessity and expense of constructing bastions was required. A system of fortifications using redans produced a series of serrations, the distance between the redans should not exceed the length of musket shot, so that fire from the faces of a redan will be able to cross the salients of the neighbouring redans.

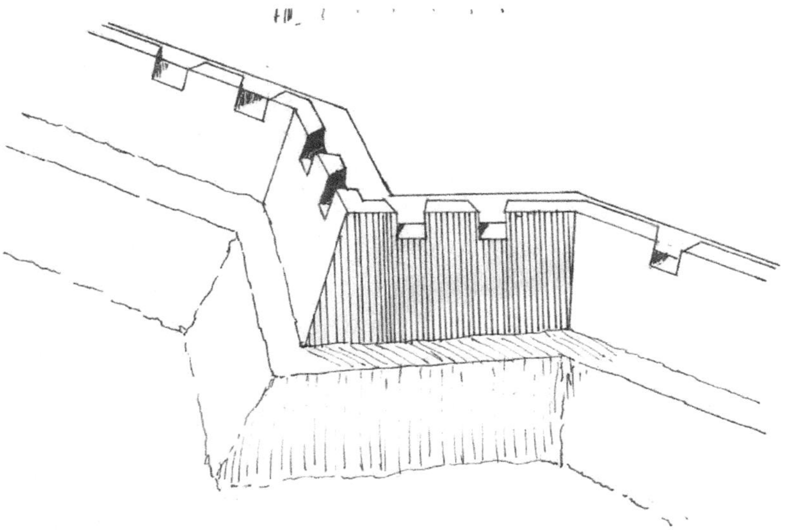

Redoubt, Redout, Reduit.
(1) A small work placed beyond the glacis, but within musket shot of the covert way, made in various forms, known as a detached redoubt.
(2) A small work built in a bastion or ravelin of a permanent fortification.
(3) An outwork or fieldwork, square or polygonal in shape without bastion or other flanking defences, sited at a distance from the main fortification, used to guard a pass or to impede the approach of an enemy force.

Re-entrant. An inward facing angle, opposite to a salient.

Re-entering angle. An angle formed by a line of fortification with its apex turned away from the besiegers and towards the centre of the place.

Re-entering place of arms. An assembly area for troops and munitions in the space left by the re-entrant angle of the counterscarp of the ditch, used to break up an enemy's initial attack or for a sortie against the enemy's lines. Dating from the late 16th century CE.

See '*place of arms*'.

Regelbau. (G. *standard design*). A book of over 600 approved plans for German bunkers and casemates of WWII.

See 'Arthaśāstra'.

Regelbauten. (G.). Standard design of German fortifications.

Régions Fortifiés (RF). (Fr. *fortified regions*). The French divided up large areas as part of the '*Maginot Line*', these regions were subdivided in the '*secteurs fortifies*' (Fr. *fortified sectors*).

See 'Maginot Line'.

Reich kasemate. (G. *realm casemate*). A casemate with small walls that jutted out which protected the embrasure, often used on the front caponier in 19th century CE polygonal forts.

See 'casemate', 'polygonal forts'.

Reinhard Line. One of the German WWII defensive lines built by Organisat Todt in Italy to slow the Allied advance, situated South of Rome, 120 km (75 miles) long, provided with light field works, and was meant to delay the allies. Also known as the *'Bernhardt Line'*.

See *'gothic', 'gustav and hitler lines'*.

Reinforced order. A bastioned trace with inset section of the curtain between bastions, to increase flanking fire, dating to the 16th century CE.

See 'bastion trace'.

Relais. An area at the base of a rampart and the top of the ditch, which was used to collect any earth which eroded from the face of the rampart.

Relief, To be relieved.

(1) To be supplied with reinforcements and supplies by one's own forces during a siege.

(2) To have a siege broken by external attack to those besieging your position.

(3) The height of a work. High relief refers to a commanding work. Low relief refers to a work that is low in height.

Relieving arches. A 13th century CE development in the masonry of curtain walls, consisting of wide relieving arches which were keyed into the towers on either side. The arches were intended to counteract the effect of sapping, for even if the base of the wall was destroyed, the wall itself would still be supported by the relieving arches.

Remblai. (Fr. *remblayer*, embank). The earth which was used to construct a parapet, rampart, etc., of a fortification, which was taken from the ditch when it was excavated.

Remise. A Dutch bombproof masonry shelter for artillery, either sunk in the ground or with earth piled on top, dating from the 19th century CE.

Remparer. (Fr. *to repair*). To repair a fortification by providing a new obstacle, or to repair a wall or ditch, with earth or wood (fascines or gabions) as a revetment to resist cannon fire.

Reirata. A temporary earthen rampart, which was raised behind a breach in a main wall, which was used to seal a breach.

Renketsukei. (J.). A Japanese castle dating from around 1600 CE, which was formed of three defensive compounds arranged in series along an axis, and connected by gateways. The three compounds were known as; the '*honmaru*', the '*ninomaru*' and the '*sannomaru*'.

See '*dôshinén*', '*teikakukei*'.

Renketsushiki tenshu. (J.). The most ornate and complex type of '*tenshu*' or principal tower of a Japanese castle, the '*tenshu*' was connected to smaller towers by parapets ('*hashidai*') at the corners of the rectangular stone base.

See '*borogata*', '*boroshiki tenshu*', 'dokuritzushiki', '*fukugôshiki tenshu*', '*reritsushiki tenshu*', *teiritsushiki tenshu*'.

Renritsushiki tenshu. (J. *connected tower*). A principal '*tenshu*' or tower of a Japanese castle which was attached to a secondary tower by a connecting parapet ('*hashidai*'). Also known as a '*renketsushiki tenshu*'. The entrance is via a '*watariyagura*' providing extra defence. Outer defences included gun ports ('*teppou sama*' and stone drops ('*ishiotoshi*').

See '*borogata*', '*boroshiki tenshu*', 'dokuritzushiki', '*fukugôshiki tenshu*', '*renketsushiki tenshu*', *teiritsushiki tenshu*'.

Retiracle. A retrenchment consisting of two faces which form a re-entering angle.

Retirade. An alteration to the face of a long bastion large enough to enable a few troops to fire on an enemy trying to gain the re-entrant angle of the ditch.

Retirata.

(1) A work built behind a bastion so as to add to its strength.

(2) An earthwork rampart, raised behind a breach, to enable the garrison to defend the position against storming.

An example is when the Turks were attacking Vienna in 1529 CE, The Count of Salm and his 16,000 defenders built a '*retirata*' behind a breach in the Southern Sector, thus preventing the Turks taking the city.

See '*retrenchment*'.

Retired flank. A flank set back or turned inwards to the rear of a work, such as a bastion where the shoulder or '*orillion*' protects the guns situated in the flank which were used to enfilade along the curtain.

See '*flank*'.

Retreat. During a siege it was an option to retreat when attacked or threatened by a relieving force.

Retrench. To construct an inner line of defence, usually consisting of a trench and a parapet.

Retrenchment.

(1)An inner line of defence, often built behind a breach.

See 'retirata' *(2)*.

(2) An inner trench and parapet provided against the loss of the outer defences.

Reveal. An opening in a parapet which has the same purpose as a crenel.

See *'crenel', 'embrasure'*.

Reverse fire. Cannon fire directed over the near side of a work from the opposite site or to the rear.

See *'fire'*.

Reverse fire gallery. A '*counterscarp gallery*' with loopholes.

Revet. To provide revetment.

Revetment.

(1) A strong retaining wall constructed on the outside of a fortification's earthwork rampart and parapet, so as to prevent it falling into the ditch.

(2) The covering of an earthen rampart to prevent it from deteriorating, using such materials as; stones, masonry, concrete or faggots.

See *'demi-revetment', 'gabion revetment' 'gazion', 'rampart', 'sod revetment'*.

Revetted earthen rampart. the first of this type was made by piling the earth excavated from the ditch behind a wooden palisade, giving the rampart a steep vertical face; the second of this type were the earthen ramparts of the 17th and 18th century CE fortresses which were revetted with stone, masonry, concrete or faggots, the ramparts were erected to offset the effect of artillery by absorbing the impact of the cannon balls. The revetment was used to hold the earth in place.

See '*rampart*'.

Riband. A thick plank or log nailed horizontally across part of a palisade to provide extra strength, attached at the base and/or just below the pointed top of the stakes of the palisade.

***See 'palisade'*.**

Rice Blockhouse. One of the types of blockhouse used in the Boer War by the English to control territory and protect the railway lines, consisting of a circle of corrugated walls with rifle ports and a gabled roof made of the same material, in some cases protected by a surrounding dry stone wall. This type of block house was designed by Major SR Rice of the Royal Engineers, and was cost effective and able to be mass produced.

See '*blockhouse*', '*Boer war blockhouse*'.

Ricetto. (It.). A small fortified area of an Italian medieval village used to store produce and livestock in case of attack. Enclosed by thick pentagonal walls and could include a guard tower.

Ricochet fire. Type of cannon fire used to attack troops sheltering behind ramparts or covered ways.

See '*fire*'.

Rideau. (Old Fr. *curtain*). A small trench protecting besiegers working close to a place.

Ridge. The crest of the glacis.

Ridge castle. See '*kammburg*'.

Rifle loop. Similar to an arrow slit but for rifles.

Rifle pit. A small work, usually in front of the main work, consisting of a short trench with a low earth rampart in front.

See '*fox hole*', '*tobruk*'.

Rimple, Georg. (c. 1636 – 1683 CE). A German military engineer, served in various European armies, then was appointed Chief Engineer for the Holy Roman Empire in 1681 CE.

Ring forts (Viking). The five ring forts identified consist of circular earthen rampart with roads connecting the four gates in the cardinal points, each of the four quarters has a long house, in one case there are four longhouses in a square for each quandrant. Located at Aggersborg, Fyrkat, Nonnebakken, Trelleborg and Borgring in Denmark, there are also sites in Sweden and northern Europe with similar construction. The ring forts were built between 970 and 980 CE during the reign of King Harald '*Bluetooth*' Gormsson and are located near important land and sea routes.

Ring of Iron. The castles in Wales commissioned by Edward 1st of England, design by James of Saint George to control the Welsh, with multiple lines of defence, some concentric in layout, with strongly built gatehouses and resupply by water, such as Rhuddlan, Flint, Caernarfon, Conway and Beaumaris.

See '*concentric castle*', '*great gatehouse*'.

Ringstand aus Stahibeton. (G.). A ring stand made of reinforced concrete for a machine gun.

Ringstände. (G. *ring stand*). Open emplacement for one or two soldiers, consisting of a circular concrete fox hole or '*tobruk*' built along the '*Atlantic wall*'.

See '*tobruk*'.

Ring wall. A stone wall which replaced the timber palisade surrounding the summit of a motte of a '*motte and bailey castle*'. Also known as a '*shell keep*'.

Ringwork. An earthen defensive enclosure, regardless of period, size or function.

See '*ring forts (Viking)*'.

Risban. A fortification used for the defence of a port or harbour.

Rivierkazemat. (Du. *river casemate*). A Dutch casemate built in the 1930's CE to defend bridges over canals and rivers, which could be several stories high.

Type A. Usually sited behind river dykes or embankments, defended by machine guns or small anti-tank guns.

Type B. Construction to take heavier fire and provided wider fields of fire than type A. Usually built where rivers offered enemy forces rapid access to the country.

See 'casemate'.

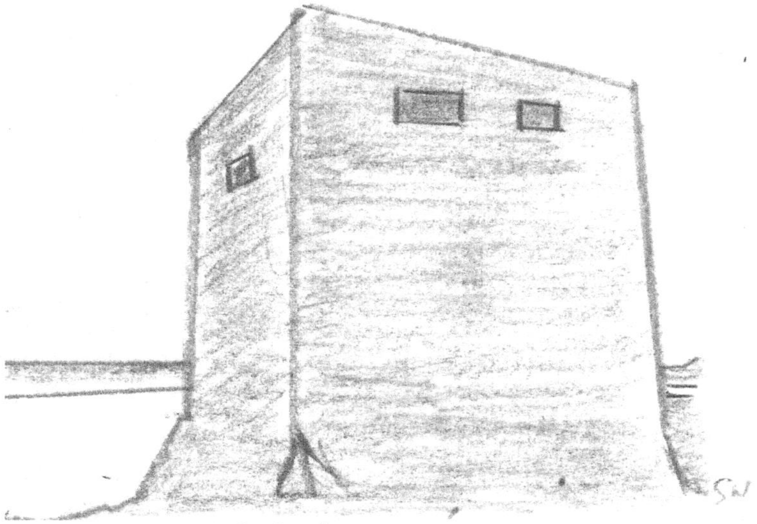

Rivierkazemate, Lent, Oost, Netherlands.

Rocade. (Fr. *ring road*). A communication channel for soldiers and materials sited behind the front line.

Rocco. An Italian castle or fort which served primarily a military purpose but could also be a residence.

See *'castello'*.

Rokudan-heki. (J.). A six tiered ishigaki wall of a Japanese castle. An example is Iwamura Castle, Mino Province in Japan.

Roll top merlon. A merlon which was designed originally to stop arrows and other missiles glancing of the merlon and passing over the parapet. The design was later used to decorate castellated buildings of the 19th century CE.

See *'merlon'*.

Rolling bridge. ***See 'Guthrie rolling bridge'.***

Rommel's Asparagus. The Allies name for poles planted in open spaces in France by the Germans to prevent gliders landing. This was done in early 1944 CE when Rommel began reinforcing the Atlantic Wall, during WWII.

See 'obstacle', 'holzpfählen'.

Rondel. A round tower which could be raised at the foot of a bastion.

Round bastion. A wide diameter low lying round bastion which was used in the Tudor artillery forts, provided with gun ports and hand gun embrasures, but dead ground was a problem in front of each bastion.

See '*artillery fortress*'.

Roundel. An artillery work built on a circular plan to height of the adjacent walls, built from earth and brickwork, later provided with casemates, first appeared in the 15th century CE, used for the use of heavy cannon,

Round gun turret. An outwork built on a circular plan and provided with embrasures, situated in the ditch or moat, dating from the 15th century CE.

Roundway. A pathway, about 3m (10ft) wide, sited below a parapet and onto the escarp wall.

See *'berm', 'chemin des rondes'.*

Royal (also Reale). A fortress which has heavy artillery mounted, or is a very strong or impressive work.

Royal military canal. England (1805 CE) A physical barrier built as a retrenchment to stop an invading army, running from Hythe to the river Rother at Rye, which would isolate the Romney Marsh, The canal itself could be used to transport troops and supplies, and provide positions for gun emplacements.

Royal Ordnance Factories (ROF) Type Pillbox. A British WWII pillbox used to defend Royal Ordnance Factories and depots, used for smaller sites, square in plan with flat roofs and long narrow embrasures on three sides and a door on the fourth side protected by a blast wall. There are 30 ROF's extant, with an example of ROF Steeton, West Yorkshire, England.

See 'British hardened field defences', 'FW3', 'pillbox'.

Ruck pillbox. A British pillbox made from the prefabricated segments of a '*Stanton air raid shelter*' submerged into the ground and re-enforced with sandbags and earth, used along the southern coast of England. Very few are extant, an example can be found at Lawyer's creek, Holbeach, Lincolnshire, England. Also known as a '*Ruck pillbox*'.

See 'British hardened field defences', 'FW3', 'pillbox'.'Stanton air raid shelter'.

Rund walle. (G. *ringwork*).

Rupnik Line. (Slovene, *Rupnikova linija*). A line of WWII fortifications built by Yugoslavia along the north and western border in response to the Italians building their Alpine wall and the threat of Germany. The line was divided into six sectors and then each sector had two lines of defence, 1) the front line made up of machinegun bunkers of various shapes and sizes, and then 2) an artillery line with anti-tank, barrier, ridge and casemate fortifications, mainly on the ridges.

See '*Alpine wall*'.

Russian sap. A shallow tunnel or covered trench dug forward of the front line to attack the enemy's trenches.

See '*sap*'.

Russia-Urkraine wall. A fortified border with the recently (2014 CE) annex Crimea consisting of trenches, dugouts, barbed wire, watch towers and surveillance systems but there were problems with funds being stolen and ongoing conflict with Russia. Construction work on the wall was stopped when Russia invaded Ukraine in February 2022 CE.

See 'linear fortifications'.

Rutba (Marsad). Small isolated fortification along the main roads for military traffic of Andalusia.

Ryousode masugata. (J.). When there are square projections on both sides of a castle entrance ('*koguchi*') for defensive purposes.

See '*masugata*', '*sodemasugata*'.

S.

Saignor. A drainage channel used to empty a ditch.

Sakamogi. (J.). A defensive fence or wall constructed on the top of an earthen wall made from thorny branches.

Saku. (J.).

(1) A small temporary palisaded fortification made from timber, logs, bamboo, narrow boards or stone pillars. The uprights were driven into the ground close enough to stop a person fitting through with cross beams at the top and bottom. The '*saku*' was used by the Yamamoto Court of Japan, in their attempts to pacify the '*Ezo*'. The '*saku*' were used in conjunction with the more substantial '*ki*' or stockade in the subjugation of the territory gained. The '*saku*' were easily erected and served as bases for further incursions into the Ezo's territory, dating from the 7th to the 9th century CE.

See '*palisade*'.

(2) A small castle. **See '*toride*'.**

Saliant, Salient.

(1) The forward projecting point of a bastion or other projecting work.

(2) A trench-system projecting towards the enemy. Salient was often used to refer to the '*Ypres Salient*' (late 1914 to 1918 CE).

Salient.

Salient (blunt).

Salient angle. The angle facing away from the centre of the place.
Salient work. A work incorporating salient angles.

See '*bastion trace*', 'redoubt', '*sconce*'.

Sally port. A small, heavily fortified gateway or gate, from the inner works to the outer works of a fortification. Used by defenders to launch sorties to take the besiegers off guard and thus gain an advantage; once the objective of the operation had been accomplished or when things were going ill for the sallying force, the defenders could fall back through the sally port before the enemy could take the advantage.
Also known as a postern. Another purpose occurred when the sally port or postern was situated in the main gate itself, this gate allowed pedestrian traffic to pass through while keeping the main gate closed.

See '*guichet*', '*postern*', '*wicket*'.

Sama. (J.). Circular, hexagonal, pentagonal, square or triangular shaped apertures in the walls of a castle for loosing arrows ('*yasama*'), or firing muskets ('*tepposama*', '*teppouzama*') or cannon ('*taihosama*') through. Many of the '*sama*' had a bigger splay on the inside than the outside to increase the field of fire while reducing the exposure of the soldier inside to return fire. '*Sama*' could also be provided with hinged wooden covers. Also known as '*hasama*'.

See 'embrasure', 'hako hazama', 'ishisama', 'shinogi hazama', 'taihosama', 'tepposama', 'yasama', 'yazama', 'zama'.

Samakubari. (J.). The prescribed distance between loopholes and the proportion of archery loops compared to gunloops based on a certain doctrine.
Samghati. (Sanskrit). A long wooden pole used to set fire to towers.
Sand bags. Cloth ('*hessian*') bags filled with sand or dirt, used to strengthen various positions including trench parapets.

See '*sand bag loophole*', 'sangar'.

Sand bag loophole. A loophole for a rifleman made from stacked sandbags on top of a parapet.

See '*loophole*', '*sandbag*'.

Sangar. (Persian, *sang*, stone). A sandbagged fire position built above ground, used when digging into the ground is not available.

See '*sandbag*'.

Sanju yagura. (J.). A three tiered turret. **See '*yagura*'.**

Sannomaru. (J.). The tertiary defensive courtyard of a Japanese castle, depending on the design, the '*sannomaru*' either surrounded the '*ninomaru*' or was sited adjacent to it.

See '*dôshinén*', '*honmaru*', '*ninomaru*',' *renketsukei*', '*teikakukei*'.

Santa Margherita Lines. (Maltese. *Is-Swar ta' Santa Firenzuola lines'. Margherita)*. A line of fortifications built in the 17th and 18th centuries CE to protect the cities of Birgu and Senglea. Also known as the '*Firenzuola lines*'.

See '*Corradino lines*', '*Cottonera lines*', '*Floriana lines*'.

Sap.

(1) A siege trench, or '*simple sap*' dug by '*sappers*' towards the enemy trenches or works.

(2) To construct trenches which were used by besiegers to approach a fortification while under cover of the trench.

(3) To undermine; to destabilise or open a breach in a wall, rampart or tower of a fortification by undermining their foundations.

(4) A communication trench between lines of trenches.

See '*trench*'.

Sap and parallel trenches. Used by a besieging force to provide protection to attacking troops from cannon and musket fire. Sappers would wheel a protective cart or '*gabion*' forward, excavate the trench and then move the cart or '*gabion*' forward again. At regular intervals parallel trenches were dug to provide connection to other sap trenches.

See 'sap', '*zig zag trench*'.

Sap head. The end of a sap, which could then be dug side ways to form a parallel or come to the surface to form an exit.

***See* '*approaches*', '*sap*', '*trench*', '*zig zag*'.**

Sappers. Soldiers used to dig mines and saps.

See '*sap*'.

Sap roller. A large '*gabion*' rolled in front of a sapper digging a trench which would act as a moving shield from enemy fire.

See '*gabion*', '*sap*', '*sappers*', '*trench*'.

Saucisson. ***See* '*battery fascine*', '*fascine*'.**

Satojiro. (J.). A village castle for a low lying position with defensive elements.

See *'jiro'*.

Sauveté. (Fr. *rescued*). Fortified villages dating to the 11th and 12th centuries CE located in southern France. Also known as *'salvetat'* or *'sauveterre'*.

See *'bastide'*.

Saxon shore fort. A part of the Roman coastal defence system from the 3rd century CE, consisting of a chain of Roman auxiliary forts (between 200 to 300 forts) sited on strategic points such as harbours and river mouths, built on both sides of the English Channel to guard the coastlines, providing both safe anchorages for the defending fleet and acting as a deterrent to raiders. Under the control of the *'Count of the Saxon Shore'*, using some existing forts at Reculver (Thames Estuary) and Branchester (The Wash). Consisting of walled enclosures with watch towers, barracks, stores, headquarters, temple, bath house and acted as shelter for the local populace during attacks. Another theory for their existence is that Carausius, a Roman naval officer rebelled against the Emperor, and he raised the forts to protect his territory.

S-Boot-Bunkers. (G. *schnellboot*, speedboat). German WWII bunker for a *'Schellboot'* (fast boat as in a torpedo moat boat) built in main harbours Holland, Belgium and France.

See 'U-boot pens'.

Scalae. (L. *ladder*). Used by attackers to scale town walls with shields over their heads.

See *'escalade'*. *'scaling ladder'*.

Scaling ladder. A ladder used in an escalade for soldiers to climb to reach the wall top and engage the opposition. The ladders had to reach the wall head and take the weight of those climbing the ladder. Usually made out of wood but could be made from leather or rope. According to one source the Turks attacking Constantinople in 1453 CE supposedly had more than two thousand ladders for a storming event.

See *'escalade'*, *'siege ladder'*.

Scarp. The side of the ditch next to the parapet; also known as the escarp.

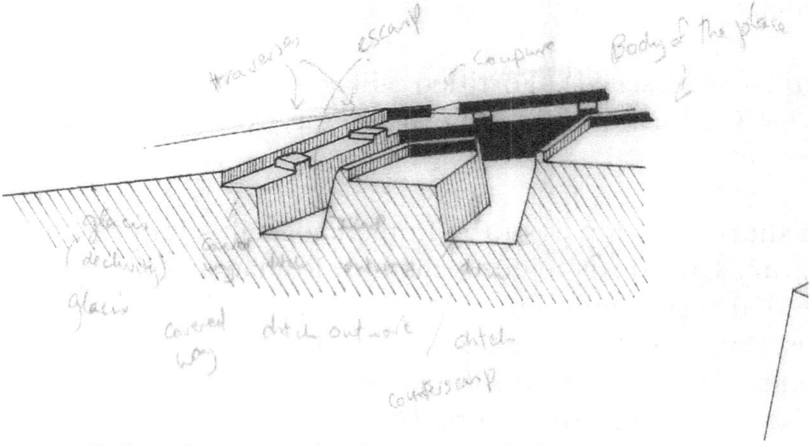

Scarped round bastion. A sloping rounded bastion provided with embrasures, dating from the 15th century CE.

Scarped wall. A wall of a fortification dating from the 14th century CE, which was developed to offset the effect of artillery on normal walls. Walls were also thickened as well as heavily scarped, in the hope that the shot would be deflected off the oblique angle that the wall presented.

Scarping. Scarping involves the increase on the angle of a slope by excavating a ditch and throwing the material downwards.

Schanze. (G. *ski jump*). A temporary defensive field work, used to block a valley or a pass, often built from *'gabions'*. Also known as a *'schanzkörbe' (G. 'bulwarks')*. A fortified line of *'schanzen'* was known as a *'vershanzte linie' (G. 'entrenched line')*.

Scharten. (G. *notches*). German term used for an embrasure or gun port.

Schartenstände. (G. *notched stand*). A casemate or bunker armed with a machine gun, anti tank gun or a cannon.

Schartenturm. (G. *notch tower*). A German WWII machine gun turret or cupola with one or more embrasures.

Schießscharten. (G.). **See *'embrasure'*.**

Schienensprerre. (G. *railway barrier).* A barrier made of embedded train rails.

See 'obstacle'.

Schildmauer. (G. *shield wall*). The especially strong wall provided with galleries and arrow slits, built across the only line of approach of a castle built on a mountain or spur; normally associated with German speaking countries.

See ***'clausurae', 'cross wall', 'hohenburg', 'shieldwall'.***

Schloss. (G.). A castle.

Schrapell mine, S-mine. An anti-personnel mine developed in the 1930's CE. When triggered, it would launch into the air, then detonate at waist height, releasing steel balls and fragments. Also known as *'bouncing betty'* or *'bounding mine'*.

See ***'mine'.***

Schu-mine 42. An anti-personnel mine consisting of just a wooden box with a hinged lid, which was cheap to make and detect, used by the Germans in WWII.

See ***'mine'.***

Schussfeld. (G. *field of fire*). ***See 'arc of fire', 'column of fire',*** **'wirkungbereich'.**

Schützenlock. (G. *protect lock*). A field work consisting of a pit for two riflemen, machine gun or mortar, which could form a strong point for a permanent position.

See 'foxhole'.

Sconce. (Du. *shans*, brushwood).

(1) A small fort located some distance from the main fortification. Either rectangular or polygonal in shape and bastioned at each corner, so that the faces between were covered by fire from flanks, giving the maximum field of fire all round defence.

See 'ensconce', 'inskonce', ***'pentagonal sconce'.***

(2) Small arches or projecting courses of stone formed across the angles of a tower. Also known as a *'squinch'*.

Screen wall, shield wall. A curtain wall or *'schildmauer'*.

See ***'schildmauer', 'shield wall'.***

Screw pickets. Twisted steel stakes (2.4 metres, 8 feet long) used to secure barbed wire to the ground, introduced in 1915 CE, WWI. The French call the stakes *'queue de cochon'* (pigtail). The bottom of the picket was bent into a spiral and the top had loops for a handle to aid in screwing the picket holder in place.

Sea fort. Coastal defence forts built to block strategic approaches to vital areas of territory (Rivers and Ports), used by numerous countries during WWII. In the Thames Estuary, UK, consisting of seven tubular steel towers connected above the water level by a lattice of steel bridges. Each tower consisted of a two storey steel structure for accommodation, various gun positions and searchlights. Used to break up enemy aircraft formation attacking London or Liverpool, prevent mine laying operations in the estuary, and prevent raids on shipping.

Sechsschartenturm. (G. *six sharp tower*). A turret with six embrasures with mountings for two machine guns and a periscope.

See '*cloche*'.

Section post, Seagull trench. British WWII hardened trench works without covers and a large number of embrasures facing the likely direction of attack. A '*section post*' was straight whereas a '*seagull trench*' was built in a '*W*' plan, mainly used for airfields. There are twenty six examples of '*section posts*' extant, and Twenty one '*seagull trenches*' extant. An example of a '*seagull trench*' can be found at the former RAF (Royal Air Force) base at Stoke Orchard, Gloucestershire, England.

See 'British hardened field defences', 'FW3', 'pillbox'.

Seirou. (J.). A watch tower, early versions were constructed of a wooden frame with no roof, later versions had a tiled roof and thick walls. '*Seirou*' could be set up in or outside a castle's walls, or on top of a castle's stone walls. '*Seirou*' varied from one to five stories in height. '*Seirou*' could be mobile structures, either on wheels ('*kurumaseirou*') or moved and erected on a new site ('*kumiage seirou*'). Also known as '*seirouyagura*'.

Seiroyagura. (J.). One of the earliest forms of '*yagura*', made from wooden beams in the shape of a square, used as watchtowers, including on the battlefield to give commanders a better view.

See '*yagura*'.

Sekimon. (J.). A castle gate made mostly of stone, examples are found in the Okinawa prefecture of Japan.

See '*mon*'.

Sekizou no mon. (J.). Castle gates with arched entrances, with single or triple arches made from stone, from the Okinawa and Kyushuu prefectures of Japan.

See '*mon*', '*sekimon*'.

Semibastion. A half bastion, or half of a bastion, obtained by dividing the bastion in two at the salient angle.

See '*bastion*', '*demibastion*'.

Séré de Rivières system. (Fr.). An array of fortifications built from 1874 CE along the frontiers and coasts of France, using polygons rather than bastions, with surrounding ditches provided with '*caponiers*', with two defensive curtains, with outer and inner line of forts, built in masonry, designed to shelter the garrison against bombardment, with artillery set out on top of the fort. Obsolete by the beginning of WWII.

See 'Barrière de fer', 'polygonal fortification', 'system of fortification'.

Serpen. An articulated ladder used for escalades.

See '*escalade*'.

Servian wall. A defensive wall around Rome dating to the 6th century BCE, 3.6 metres (12 feet) wide, with a berm 6 metres (20 feet) wide, with a ditch 30 metres (98 feet) wide and 9 metres (30 feet) deep. Provided with several gates which were guarded by internal towers.

See '***Aurelian wall***'.

Serving room. A room usually in a coastal battery where cartridges were issued from.

S-4 Kazemat. The smallest type of Dutch re-enforced concrete casemate, a 1930's CE part of the pre-WWII defensive program, defended by a light machine gun, often defending canal bridges or minor roadways.

See 'casemate'.

Shell filling room. A bomb proof casemate or building for loading shells and cartridges with gunpowder.

Shell keep. The '*motte and bailey castles*' of the Normans served their purpose well in the subjugation of England but wood does burn and rot, and stone lasts much longer. The use of stone in their constructions was gradually introduced, starting with chapel, or the gate which was a chief source of concern. By the beginning of the 12th century CE, wooden palisades surrounding the top of the motte had started to be replaced by a stone wall, thus making what is known as a shell keep. The surrounding wall was provided with a sentry walk around its perimeter, and the domestic buildings normally housed in the donjon were attached to the inside of this wall, leaving space for a courtyard in the centre.

See '*keep*', '*motte*', '*ring wall*'.

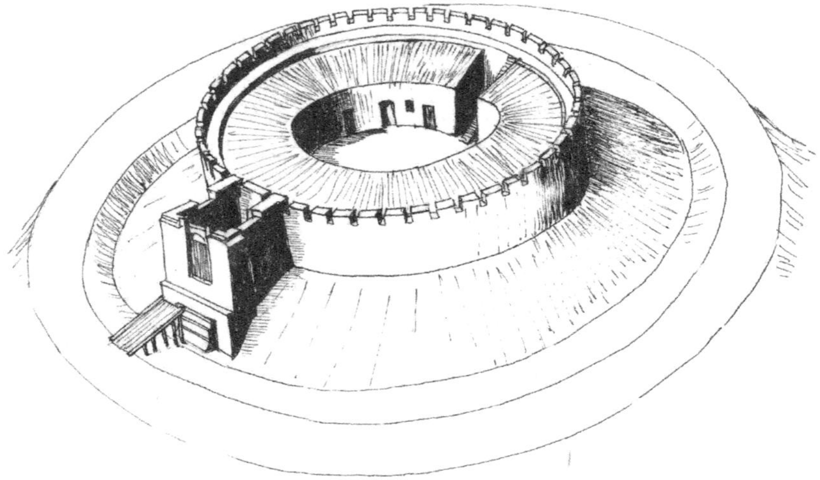

Shell keep. Restormel, Cornwall, England.

See *'Appendix 1. What is a Castle?'*.

Shell scrape. A shallow depression that only offers protection on the prone position.

Shelter trench. A simple rifleman's trench with a rear ditch.

See *'trench'*.

Shibadoi. (J.). A grass covered earthen embankment (*'doi'*) with inward sloping sides and flat top, used near main entrances (*'koguchi'*) or at the base of towers (*'yagura'*).

See *'doi'*.

Shifting house. A building where gunpowder was prepared and checked.

See 'arsenal'.

Shield wall. A stone wall built to protect a pass or fort (i.e. the Hellenistic screen walls of the Isthmus of Corinth (Greece)).

See *'clausurae'*, *'cross wall'*, *'schildmauer'*.

Shikirimon. (J. *partition gate*). A gate in a Japanese castle that partitions off an area of the castle.

See *'mon'*.

Shinogi hazama. (J. *hazama,* to fire a gun or loose an arrow). A triangular loophole.

See 'hako hazama', 'ishisama', 'sama', *'taihosama'*, *'tepposama'*, *'yasama'*, *'zama'*.

Shiro. (J.). A Japanese castle or fortification.

See *'jo"*.

See '*Appendix 2. Shiro - A Japanese Castle?*'.

Shirodori nawabari. (J.). The process of castle site selection, layout of the compounds and the construction of the castle itself.

See '*kuruwakubari*', '*nawabari*', '*samakubari*'.

Shi ta men. (Ch. *solid bed door*). A ledged gate door, with the ledges directly slotted in the boards of the door with sliding dovetail joints, and mortice and tenon joints on the sides, and all swings on door pivots.

Shitomi. (J.). Items such as walls, embankments, even plants to prevent an enemy observing inside a castle main gate, between two compounds or where pathways made sharp turns.

See 'kazashi'.

Shi xin di tai. *See 'di tai'.*

Shot deflecting battlement. A battlement where the merlons have a sloping apex which inclines inwards, designed to deflect cannonballs.

See 'curved merlon', 'merlon'.

Shojibori. (J.). A wide trench with square depressions carved out. A good example is at Yamanaka Castle in Shizuoka.

See 'bori', 'hakobori', '*horikiri*', '*kenukibori*', '*tatebori*', '*unebori*', '*unejotatebori*', '*yagenbori*'.

Shot hole. An opening in a fortification for firing firearms through, usually smaller than a gun port or an embrasure.

Shot trap. See '*faussebraye*'.

Shoulder angle. The angle made by the intersection of the flank and face of a bastion.

Shoulder of the bastion. The angled part of a bastion between its face and its flank, normally its widest point, the shoulder was sometimes supported by the '*orillion*'.

Shoulder angle. The angle formed between the face and the flank of a '*bastion*' or a '*demibastion*'.

Shukuruwa. (J.). The main bailey of a castle.

See '*kuruwa*'.

Shurrafa. (Ar. *crenulations*). **See '*crenel*'.**

Siege. (L. *sedes*, seat). A military operation carried out by an armed force whose purpose it was to gain entry and control of a position or fortification, by cutting off access to drinking water, blocking the supply lines and escape routes and forcing entry.

See '*Appendix 3. Siege Warfare - The art of Offence and Defence*'.

Siege castle. A castle erected by a besieging force close to an invested fortress, to cover a gate or postern. The siege castle was used to prevent the defenders. launching sorties, receiving revenues and the produce from the surrounding area; and also preventing relieving forces and communications reaching the besieged. From the siege castle the attackers could launch raids, dominate the roads and surrounding area. A castle survives off its surrounding lands and trade, deny these and the occupants of the castle have to live on what they have stored.

See *'counter castle', 'counter work', 'malvision'.*

Siege engines. Chroniclers of the time used a range of terms for siege engines which are hard to interpret today, they include: Algarrades, Brigoles, Calabres, Chaabla, Fonevols, Fundae, Mangonas, Paterells and Springalds.

See *'mangonel', 'trebuchet'.*
See *'Appendix 3. Siege Warfare - The art of Offence and Defence'.*

Siege ladder, Ms. Germ. Qu. 15 Folio 72r (1460 CE).

Siege lines. The lines of works and fortifications built by the besiegers and the besieged, to strengthen their positions during a siege.

See 'siege'.

Siege park. A place for the collection and storage of entrenching tools, explosives and other equipment before a siege begins.

See *'artillery park'*.

Siege tower. Mobile multi-tiered tower, made of timber, could be covered in fresh animals to resist fire, rolled up to the walls of a castle on wheels or rollers, where soldiers could cross a bridge laid down from the top of the tower. The lower story could also be furnished with a battering ram to attack the base of the wall.

See *'siege'*, *'turres ambulatoriae'*.

Siege train. A wagon train carrying the siege equipment, provisions and other supplies for conducting a siege operation.

Siege works. The works erected by besiegers for either offence or defence.

See *'facines'*, *'field work'*, *'gabion'*, *'line of circumvallation'*, *'malvoisin'*, *'parallel'*, *'sap'*.

Siegfried. A German WWII field shelter made from earth and sheet metal.

Siegfried line. A defence line built in the late 1930's CE and re-enforced in the mid 1940's CE by the Germans (WWII) to defend Germany from attack from the West, running from the Netherlands to Switzerland (630 km, 390 miles), using dragon's teeth, bunkers, tank traps and machine gun emplacements, which failed to stop the advance of the allied forces in 1945 CE. Also known as '*Scharnhorst*' or '*West Wall*'.

Signal station. A site which was on a prominent hill top and had commanding views of the surrounding countryside, which could communicate with other signal stations using semaphore. Other sites could be used with signal towers or strategically located buildings.

Silesian walls. (Polish, *Wały, Slaskie*). Lines of parallel ramparts and ditches running 30 km (19 miles) running through Lower Silesia in Poland , by the towns Szprotawa and Kożuchów forming the southwest border of the Duchy of Głogów, built between 1413 and 1467 CE.

Sillon. (L. *sulcus*, furrow). A small rampart built in front of a large moat.

Skåne Line. (Swedish, *Skånelinjen*). A double line of fortifications, consisting of a first line of concrete bunkers along the shore, armed with machine guns and light cannons. The 2nd line has armed troops 300 meters (984 feet) inland behind barbed wire, some in concrete bunkers. ,about 500 km (300 miles) long, built during 1939-1940 CE around the southern coast of Sweden, from Halland to Blekinge, to protect against German or Russian attack.

Skids. A series of logs running perpendicular to the parapet with head logs. If enemy artillery hit the head long and knocked it back it would move down the skids instead of hitting the soldiers in the trench.

See '*head log*'.

Skolops. (Gr.). Palisade stake.

Slashing. Cutting down the trees and undergrowth in front of work to create a clear field of fire. If the trees are felled towards the likely direction of attack and the tree branches sharpened to form '*abatis*'.

See '*abatis*'.

Slide. The base of a heavy gun upon which it recoils.

Slight. To damage a fortification or part of, so as to make it indefensible.

Slit trench. A narrow trench dug deep enough into the ground to offer a standing soldier some protective capacity during an attack such as artillery bombardment or air raids. Also known as a '*fire trench*'.

See '*fox hole*', '*trench*'.

Small keep. A small adjoining tower incorporating a staircase leading to the first floor entrance of a 10th or 11th century CE Norman castle. The staircase was entirely unprotected, so defenders could easily deny their use to any attackers who might try to gain entrance to the keep.

See '*forebuilding*', '*forework*', '*keep*', '*penthouse*'.

Sodekuruwa. (J. *sleeve compound*). Compounds which are sited on a mountainside and are thought to deviate from the line of the mountain ridge. '*Sodekuruwa*' are sometimes arranged in a step formation over an entire section of the mountainside or in a spiral form around the mountain but the distinction between '*koshiguruwa*' and '*sodekuruwa*' is not always clear.

See '*koshiguruwa*', '*kuruwa*'.

Sodemasugata. (J.). The square projection of a castle wall on one side of a castle entrance ('*koguchi*') for defensive purposes.

See '*masugata*', '*ryousode masugata*'.

Sod revetment. A '*revetment*' made from cut '*sod*' (turf) used to reinforce a wall, parapets, or the '*cheeks*' of embrasures, which is held in place by wooden stakes.

See '*revetment*'.

Sogamae. (J.). Defences such as earthworks or moats.

Sookarika. (Sanskrit). A bamboo mat covered in leather used to protect a tower from stone missiles.

Sorobanbashi. (J.). **See '*hikuhashi*', '*kurumabashi*'.**

Sole. The bottom or floor of an embrasure.

See 'embrasure'.

Solid bastion. A bastion is solid when the *'terre-plein'* is full of earth.

See 'bastion', 'empty bastion'.

Sotobori. (J.). The outer moat of a castle that has two or more moats.

See *'bori', 'karabori mizubori', 'yagenbori'.*

Sotogata. (J.). A multilevel *'tenshu'* which was built on a squared stone foundation, with each level the same shape as the one below it but smaller in size.

See *'borogata', 'tenshu'.*

Sotoguruwa. (J.). An outer bailey, including the topographical features such as waterways and mountains.

See *'kuruwa'.*

Sougamae. (J.). An enclosure formed by an earthen embankment, a stone wall and a moat of a castle town (*'joukamachi'*). If there is more than one such surrounding enclosure, the outermost enclosure is the *'sougamae'*.

See *'doi', 'joukamachi'. 'kamae', 'kuruwa'.*

Soutougata tenshu. (J.). An ornate multi-storied *'tenshu'*. Also known as a *'soutou tenshu'*.

See *'tenshu'.*

Spall skin. Even a thin steel skin can stop the flakes or fragments from the inside concrete surface of a bunker becoming dangerous projectile during a bombardment.

Spanish rider. A movable wire obstacle attached to a frame, used to block access.

See *'cheval de frise', 'knife rest'.*

Specula. (L. *watch*). Roman square watch towers guarding the German *'Limes'* (*defensive frontier*), with walls 3.5 - 4.5 metres (12 or 15 feet) thick.

See *'burgi', 'limes', 'vigilarium'.*

Speculartorium. (L. *speculari*, to spy out). A window or peep hole in a door or a gate, with a hatch and a grating, used to observe people before the gate.

See 'grille', *'manidwar'.*

Sperre. (G. *lock*). A barrier.

Spider hole. A Vietcong hole in the ground with camouflage cover, used to ambush enemy soldiers.

See 'camourflage'.

Spider wire entanglements. A network of single strand barbed wire fences, with sections about 30 metres (98 feet) in length, forming irregular shapes.

See 'barbed wire', 'high wire entanglements','low wire entanglements'.

Spigot mortar emplacement. A British WWII anti-tank emplacement consisted of an un-roofed earthwork, sometimes constructed in brick or concrete or revetted with corrugated iron, with a central concrete pedestal with stainless steel peg to hold the mortar. An example was excavated by Time Team (Series 15, Episode 6) at Shooters hill, London, England.

See 'British hardened field defences', 'FW3', 'pillbox', 'stop lines'.

Spina. (L. *spine*). The central partition of a double gate of a Roman fort.

Spitzbunker. See *'winkel bunker'*.

Splayed base, Splayed talus. A wall or tower was in some cases supplied with splayed talus or base. This was for a number of reasons, such as; to deflect projectiles dropped by defenders from the walls above to attack besiegers below, and to counter sapping and the effects of artillery.

See *'plinth', 'spur', 'talus'*.

Splay. The angle of the sides of an embrasure, the wider the angle, the bigger the arc of fire for the projectile.

See 'arc of fire', 'embrasure'.

Splayed embrasure. An embrasure which was splayed to enable a cannon to be traversed in a greater arc than with a conventional embrasure, giving the cannon a wider field of fire and the ability to train the fire on a moving target. Used in forts, especially those used for coastal defence.

Spike. To drive an iron spike into the touch hole of a cannon and render the cannon un-usable if captured by the enemy.

Spoila. (L.). The use of previously cut stone in the construction of later fortification, which reduced the labour and transport expenses, especially in urgent need in the face of attack. Spoila can also be used to set a date for the construction using the date of the original stone work (i.e. the Herulian Wall in Athens, Greece).

Spornburg. (G. *spur castle*). A German castle built on a hill spur surrounded on three sides by steep terrain and only having to provide defence on one side.

See 'castle'.

Spreizhbemmbalk. (G. *expansion beam*)

See 'strand beam'.

Spur.
(1) An angular projection of masonry applied to the base of a tower. The spur design had the double advantage of fending off projectiles from siege engines, and of increasing the difficulty of sapping. Such protruding spurs or prows became quite common on keeps and towers of the late 12th century CE.

See 'batter', 'plinth', 'talus'.

(2) A wall joining a rampart to an inner wall.
(3) A tower or blockhouse used to form a salient angle in the outworks.
Spur buttress. A sloping support for a wall or tower.

See *'spur (1)'*.

Spur castle. See *'spornburg'*.
Spy hole. See *'oilet'*.
Square flanking tower.
(1) A tower situated at the corner of either a rectangular or square keep, flanking the keeps walls (i.e. Rochester castle).
(2) A square tower used to flank a curtain wall.

See 'flanking tower'.

Square keep. A keep which was built on a square plan, possessing advantages over the wooden donjon design which preceded it, such as; they could not be fired, they could be built to a greater height, and because of their height they did not need the same concentration of defenders as was needed for the defence of the lower earth and timber walls.
Their disadvantages were that their square corners could easily be attacked with a bore or by undermining, and the square shape also limited the field of fire. The development of the polygonal and round keeps offset most of these faults.

See *'keep'*, *'polygonal keep'*.

Squinch. See *'sconce (2)'*.
Stacheldraht. (G. *barbed wire*). ***See 'barbed wire'***.
Staggered arrow loops. Arrow loops which were constructed in a staggered pattern to avoid excess strain on any part of a wall or creation of any weak areas.

See *'quincuncial disposed arrow slits'*.

Stahlbeton. (G.). Reinforced concrete.
Stahligen. (G. *steel hedgehog*). A German barrier made from three crossed girders.

See 'Czech hedgehog', 'obstacle'.

Staircase turret. A turret at the angle of a tower or keep which contained a staircase used to gain the parapet and provided with arrow slits which were used for flanking the defences.
Stalin line. A line of fortifications along the western border of the Soviet Union (USSR) running for 1835 km (1140 miles), building started the 1928 CE and continued till 1939 CE, consisting of bunkers and gun emplacements. Not built as a continuous line but was meant to funnel invaders along set out corridors, similar to the '*Maginot line*' but less elaborate. Most of the German army in WWII by-passed the the isolated bunkers.
Stacheldraht. (Fr.). A late 13th century CE trebuchet with a movable counterweight to improve aiming.

See 'trebuchet'.

Stalperdranht hindernis. (G. *trip wire obstacle*).
Stambha-durga. (Sanskrit). An Indian fort sited in a forest of tall trees, water sources are lacking.

See 'durga'.

Stanchion. A vertical iron bar securing a window light.
Ständige anlage. (G.). A permanent installation.
Stanton air raid shelter. Made from segmented steel reinforced concrete joined together with an arched roof, of any length manufactured by the Stanton Iron Works for the Air Ministry.

See *'Anderson hut', 'flak towers (Germany)', 'Morrison shelter', 'Winkel bunker'.*

Star fort. A field work constructed in the shape of a star, by using four or more salients and no curtains.

See *'bastion', 'bastion fort', 'bastion trace'.*

Stekelvarken. (Du. *porcupine*). A small Dutch rifle and machine gun bunker in a trapezium shape, dating from the 1930's CE, with a number of variants depending on the number of embrasures and rifle holes.

See 'bunker'.

Stellar trace. A work in the form of a multi-pointed star.

See *'trace'.*

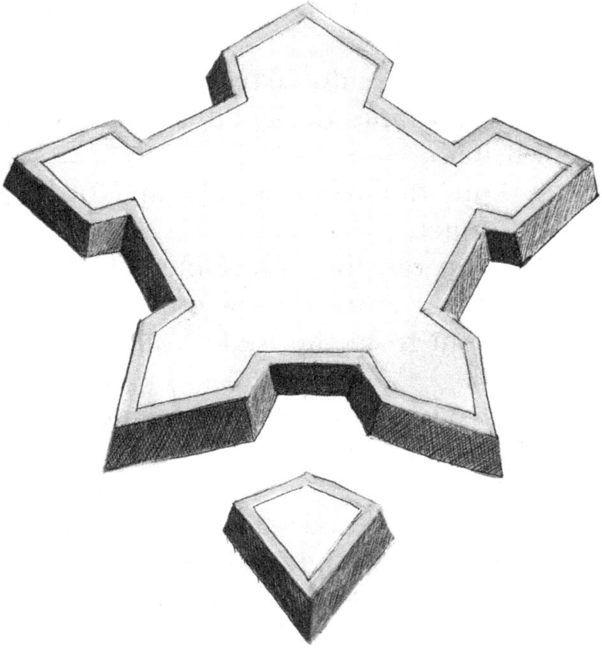

Stellung. (G.). A position which can be part of a fortified line.

Stepped merlon. A merlon which rose to its apex in two or more steps. A variation was where the top of the steps was sloped towards the exterior of the parapet.

See *'irish battlements'*, *'merlon'* *'venetian battlements'*.

Sthala-durga. (Sanskrit, *plain fort*). A fort surrounded by artificial moats with water supplied by a river.

See 'durga'.

Stimuli, Stimulus. (L.). An obstacle formed from a stake sunk into the ground with a projecting iron spike, placed forwards of a Roman defence work like at Alesia.

Stins. *See 'borg'.*

Stockade.

(1) To fortify with a breastwork.

(2) A defensive enclosure made of upright stakes.

See 'palisade'.

Stop lines. Stop lines. During WWII in 1940 and 1941 CE Britain was expecting a German invasion, more than fifty *'stop lines'* lines were constructed around the United Kingdom, consisting of pillboxes and anti-tank obstacles set up inland from the coast to stop the advance of the Germans including taking advantage of rivers and other natural obstacles, designed to compartmentalise the country and delay the Germans long enough for more mobile forces to counter-attack.
Examples of the stop lines are:
The **GHQ Line (General Headquarters Line)** the GHQ Line was the longest and most important, designed to protect London and the industrial heart of Britain;
The **Coquet Stop Line** ran from Amble in Northumberland up the valley of the River Coquet;
The **Taunton Stop Line** ran through Somerset, Dorset and Devon, to Taunton, then to the River Parrett to the coast near Highbridge;
The **Tyne Stop Line** ran from the Rede Valley in Northumberland to Newcastle upon Tyne.

See Army anti-tank island', 'linear fortifications'.

Storm. A swift attack of a fortification or part of, with overwhelming force.

Storm poles. Wooden pointed stakes (i.e. fraises) which were projected horizontally from the top of the interior slope of the ramparts of a field work. Used to impede the enemy scaling the ramparts. An Example is the Queen's Sconce at Newark, April 1646 CE.

See *'pentagonal sconce'*.

Storm-proof work. A defended position where an approaching enemy force is destroyed as soon as they are within range, thus unassailable.

Strand bars. ***See 'hemmkurven'.***

Strand beam. ***See 'beach obstacle', 'Atlantic wall', 'hemmbalk', 'obstacle'.***

Strata Diocletiana. (L. *Road of Diocletian*, 284 – 305 CE), The *'limes Arabicus'*, a fortified road along the eastern desert border of the Empire from the southern bank of the Euphrates river passing Palmyra and Damascus to northeast Arabia, defended by Roman rectangular forts for auxiliary units *'quadriburgia'* sited every 20 Roman miles (a day's march) or 1.48 km (0.92 miles).

See 'quadriburgia', 'limes'.

Strategical fortifications. Normally only constructed in time of peace to strengthen a point of strategic importance in the event of war.

See *'tactical fortification'*.

S-3 Kazemat. A type of Dutch reinforced concrete casemate, a 1930's CE part of the pre WWII defensive program, defended by a light or heavy machine gun, often sited along canals, roads, and railway lines. Often sited with other bunkers with the usual fields of fire that overlapped.

See 'casemate'.

Sturmstangen. (G, *storm poles*). **See *'storm poles'*.**

Stützpunkt (Stp). (G. *base*). A German WWII fortified points used in the *'Atlantic wall'*, using barbed wire and mines, often with artillery, the size of which depended on the site's strategic importance.

See 'Atlantic wall', 'stutzpunktgruppe'.

Stutzpunktgruppe (StPGr). (G. *base group*). A group of *'stutzpunkts'* which could cover several kilometres of shore and include army, naval and air force units, along with hospitals, stores, anti-aircraft batteries, radar stations and command posts.

See 'Atlantic wall', 'stutzpunkt'.

Switch lines. A linear fortification or field work, with one end attached to the main line of defence, so when an enemy breaks through the main line of defence they have to swing around to face this new obstacle, thus exposing their flank and rear to counter attack.

Stratopedon. (Gr. *camp*). An army camp, equivalent to the Roman *'castra'*.

Strigae. (L.). The various rectangular spaces marked out for the erection of the tents of a Roman encampment.

Stronghold. A fortress, or a fortified or defensible place.

Submarine pens. A reinforced concrete shed to protect submarines during repairs and restock of supplies from attack by enemy aircraft.

See 'U-boat pens'.

Successive portcullis. Two or more successive portcullises defending the entry of a gatehouse or barbican; each was raised by separate machinery, placed on different floors without communication between them so that if a traitor could get one open, he was not able to open the others. The enemy could also be trapped between the portcullises if the gate was forced.

See *'portcullis'*.

Sudes, Sudis (L.).

(1) A wooden stake used by soldiers to practice sword drill.

(2) A wooden stake carried by legionarii to fortify a rampart of a Roman marching camp (*'castra'*).

See *'vallus'*.

Südwall. (G. *south wall*). A German Coastal WWII fortification along the Mediterranean coast (864 km, 537 miles), consisting of bunkers and forts, dating from 1942 CE, similar to the Atlantic wall. Also known as the '*Mediterranean Wall*'.

See 'Atlantic wall'.

Suffolk square pillbox. A British WWII pillbox unique to Suffolk, square in plan, with two embrasures in front, an embrasure to each flank, and a door in the rear protected by an embrasure and a porch, 37 examples remain, including one at Benacre, East Suffolk, England.

See 'British hardened field defences', 'FW3', 'pillbox'.

Sukigamae, Sukashigamae. (J.). An embankment that does allow the view of the buildings inside the compound.

See '*doi*', '*kamae*', 'kuruwa'.

Sumijiri ishi. (J. *corner tail stone*). The stone that butts up against the rear face of the corner of a stone wall ('*ishigaki*') of a Japanese castle.

See '*ishigaki*'.

Sumiumadashi. (J.). A barrier built at the corner of an outer castle compound.

See '*umadashi*'.

Sumi yagura. (J.). A wall tower situated at the corners of the stone walls of a Japanese castle, which was used to defend the castle walls. The height of the '*sumi yagura*' ranged from one to three stories, some actually took on the exterior of a '*tenshu*'.

See '*yagura*'.

Sump. A drainage ditch or pit at the bottom of the trench used to control the amount of water present.

See '*coffer*'.

Superior slope. The surface of the parapet near the crest.

Supporting points. An area of ground or works organised defensively, which may consist of natural or constructed features (i.e. woods, villages or trenches).

See '*fortified line of resistance*'.

Support trenches. The second line of defence, in case the enemy took control of the front line trench. Also contained support to the front line including first aid stations and kitchens.

See '*trench*'.

Surovikin line. A set of fortifications built by the Russians in Ukraine in 2022 CE, about 2,000 km (1243 miles) long from Belarus to the Dnipro delta, consisting of trenches, artillery positions, anti-personnel and vehicle mines, razor wire, earthen ramparts and dragon's teeth.

See 'linear fortifications', 'Wagner line'.

Sutebori. (J.). Moats excavated in the open areas on the approaches to a castle.

See '*bori*'.

Suteguruwa. (J.). A compound with limited defences, used to launching offensive operations, overlooked by the main compound.

See '*kuruwa*'.

Svan (Svaneti). (Georgian '*svanuri k'oshk'i*', tower house). Defensive tower from Georgia, dating between the 9th and 12th centuries CE, the towers are either free standing or connected to a residence, tapering in with towards the top, three to five stories high, topped machicolations and parapets.

See '*casatorri*' in Italy, '*vainakh*' *in* Chechnya and '*vepshi*' in Ingushetia, Russia.

Switch line. A switch line was a trench that connected two trenches that ran in parallel to the front line, the same can be said of defence lines ('*Hitler and Gustav lines*').

See 'stop lines'.

Swivel pins. The hinges used in certain types of drawbridges.

System of fortification. A formalisation of fortification arrangements and proportioning, usually associated with the inventor or with the country it was first used extensively.

See '*perpendicular fortification*', '*séré de rivières system*', 'trench system', '*Vauban's three systems*'.

T.

Tabica, tabby. (Ar.). A concrete used in fortifications.

Tabor. (Slovenian). A defensive stone wall built around a church, built during the Ottoman Turk attacks in the 16th century CE.

Tachisama. (J.). Loopholes set at a height to be fired through while standing.

See *'isama'*, *'sama'*.

Tactical fortifications. Simple works constructed by troops in the field to deal with a tactical need, such as *'entrenchments'* which are constructed from materials and with tools close to hand. Also known as *'field works'*.

See 'strategic *fortification*'.

Taif. (Ar.). A 6th century CE Arabian fortification consisting of a large walled enclosure.

Taihosama. (J.). An aperture in the walls for firing cannons through.

See 'hako hazama', 'ishisama', 'sama', 'shinogi hazama', 'tepposama', 'yasama', 'zama'.

Taiko yagura. (J.). Drum turret of a castle.

See *'yagura'*.

Tai no shiro. (J. *confronting castle*). A temporary fortification used to attack a castle. Also known as *'mukaijiro'* (*facing castle*) or *'tsukejiro'* (*annex castle*).

See *'counter castle'*, *'malovoisin'*, *'shiro'*.

Talcaya. (Ar.). **See 'atalaya'.**

Talus. (L. *ankle*). The sloping or scarped face at the base of a fortification's wall. The talus reduced the effectiveness of scaling ladders in the following fashion; if the ladders were placed on the ground, they would either be too short or inclined at such an oblique angle the structural strength was severely reduced. The talus was also used by defenders to deflect stones thrown from above at the enemy below, the stones on hitting the talus splintered and ricocheted with a shrapnel effect. An example is from Montsorieu Castle, Spain.

See *'batter'*, *'plinth'*, *'spur'*.

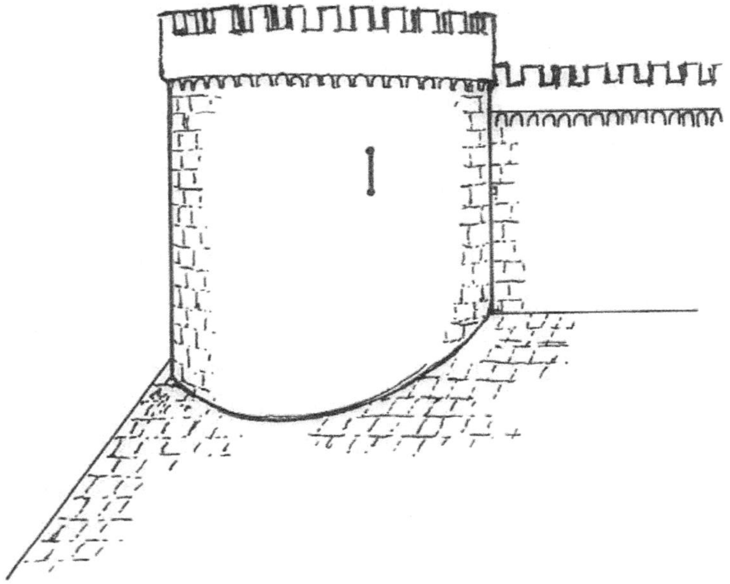

Tamaishizumi. (J.). A type of '*ishigaki*' construction using river stone.

See '*ishigaki*', 'kikkozumi', '*kirikomihagi*', '*nozurazumi*', '*nunozumi*', '*ranzumi*' '*tanizumi*','*uchikomihagi*'.

Tambour. (Fr. for a type of drum, from Persian *tambur*). A small defence work made of palisading or earth, usually in the shape of a '*redan*', used to defend gates, or other entrances.

Tamon. (J.). **See '*tamon yagura*'.**

Tamon yagura, (J. *many listener tower*). Edo period (1603-1867 CE) Japan, consisting of a continuous fortified gallery of rooms which ran along the top of the stone walls with watch towers at each end, could be between one and three stories high, providing defensive structure (provided with '*sama*' and '*ishiotoshi*'), living quarters or storage areas. Also known as '*tamon zukuri*' and '*tamon nagaya*' (J. *many gate long gallery*). An example can be found at Himeji Castle, Hyōgo prefecture.

See '*yagura*'.

Tanizumi. (J.). A type of '*ishigaki*' where the stones are laid in a diagonal pattern which aids in stability, often used in the late Edo period. Also known as '*otoshizumi*'.

See '*ishigaki*', 'kikkozumi', '*kirikomihagi*', '*nobori ishigaki*' '*nozurazumi*', '*nunozumi*', '*ranzumi*' 'tamaishizumi','*tanizumi*','*uchikomihagi*'.

Tank defilade. A tank position where the tank is hidden by natural or artificial means to hide the tank in a '*hull down*' or '*turret down*' position, where the tank can move up, fire at the enemy and return to the defilade position.

See '*defilade*'.

Tank island. *See Army anti-tank island', 'stop lines'.*

Tank shelter. An underground concrete storage area for one or two tanks with armoured doors and access ramp.

Tank traps. See '*anti-tank obstacles*'.

Tank turret. During the 20th century CE spare tank turrets were used on concrete bunkers.

See 'naval turret', 'tobruk'.

Tapia. A mixture of gravel and cement, dried between wooden boards. Used to build fortifications and other buildings in Spain.

Tapecu. (Fr.) A gate swinging on a pivot at the top like a shutter, used on posterns, in the 12th to 14th century CE, France. See the Gate of Aubeton, attacked by the Count of Hainault, MS Froissart, BNF Francais 2643 volume 1, dating from the 15th century CE.

See '*drawbridge*', '*hane-age-mon*', '*hikibashi*', '*porticulis*', '*yett*'.

Taphros. (Gr.). Ditch or moat.

Tatakidoi. (J.). A pounded earthen embankment, using multiple layers of soil mixed with water then compacted, to about 6 metres (19 feet) in height.

See '*doi*'.

Tate. (J.). A small castle with earthen embankments ('*doi*') and a moat ('*hori*'). Also known as '*tachi*'.

See '*yashikijiro*'.

Tatebori. (J.). A trench running up a slope. It can be used to impede attackers from easily crossing a slope. One side of the trench was steeper to make it hard to climb out of. Also known as a '*tatsubori*'.

See 'bori', 'hakobori', 'horikiri', 'karabori', 'kenukibori', 'shojibori', 'unebori', 'unejotatebori', 'yagenbori'.

Teicho. (Gr.). Wall.

Teikakukei. (J.). A Japanese castle design dating from around 1600 CE, consisting of two defensive compounds; the inner or '*honmaru*' was set to one side of the encircling secondary compound or '*ninomaru*'. This positioning of the '*honmaru*' was to remove it from the most likely line of assault, thus gaining extra protection.

See '*honmaru*', '*ninomaru*'.

Teikakushiki. (J.). Connection of multiple barriers (*'umadashi'*) along an axis to protect a castle's entrance (*'koguchi'*) Also known for its full name *'teikakushiki nawabari'*.

See *'koguchi'*, *'umadashi'*.

Teiritsushiki tenshu. (J.). A *'tenshu'* which consisted of the principal tower attached directly to a secondary tower, also known as a *'fukugôshiki tenshu'*.

See *'borogata'*, *'boroshiki tenshu'*, 'dokuritzushiki', *'fukugôshiki tenshu'*, *'renketsushiki tenshu'*, *'reritsushiki tenshu'*.

Teller anti-tank mine. A German anti-tank mine of WWII that came in a plate shaped metal casing with a handle, with a high operating pressure which would only be set off by a vehicle passing over it.

See *'mine'*.

Temporary fortification. A fortification constructed for a specific battle or campaign, made with the available materials, usually constructed in a single day.

***See 'temporary trenches'*.**

Temporary trenches. Such trenches would be dug to occupy newly controlled ground from a counter attack, or if an enemy had broken through the main entrenchment. The temporary trenches were only waist deep lacking the usual infrastructure including support or communication trenches, or dugouts.

See *'trench'*.

Tenaille. A small low work situated in the ditch between bastions, used to provide extra covering fire for the curtain wall, there were three types. The first had faces like a bastion but were much lower; the second had faces, flanks and curtains; and the third had only faces.

See *'redan'*.

Tenaille trace. A trace formed of redans joining at right angles to form a serrated front.

See *'redan (2)'*.

Tenaillon. A work raised on either side of a ravelin, used to add strength and cover to the shoulder of the bastion.

Teng pai. (Ch.). A Chinese defensive obstacle used to protect strategic sites, forts and cities, dating from the 4th century BCE. Also known as *'mo pai'*. According to a Ming dynasty military treatise, the obstacle was also known as *'juma'* (*horse repeller*) or *'iujiao'* (*deer horn*).

***See* 'chevaux de frise', *'hersillion'*.**

Tenshu. (J.). The main tower of a Japanese castle, constructed mainly of timber on a base consisting of *'ishigaki'* or earth revetted with dry stone walling, serving a number of purposes such as; ceremonial, residential and military. The early *'tenshu'* were simple residential buildings with belvederes perched on their rooftops, later they developed into the larger multi-levelled towers with up to five stories (*'soutougata tenshu'*). Single *'tenshu'* were called *'dokuritsu tenshu'*, large tenshu were called *'daitenshu'*. Smaller 'tenshu' called *'kotenshu'* could be attached along with connecting galleries *'watariyagura'* to form a complex compound (*'fukugou tenshu'*). *'Tenshu'* linked in a tower complex is called *'renketsu tenshu'* and a group of *'tenshus'* is called a *'renitsu tenshu'*. Also known as *'tenshukaku'* and *'tenshuyagura'*.

See *'Appendix 2. Shiro - a Japanese Castle?'*.
See *'boroshiki tenshu, 'dokuritzushiki', 'fukugôshiki tenshu', 'renketsushiki tenshu', 'reritsushiki tenshu', teiritsushiki tenshu'*.

Tenshudai (J.). The foundation of the main keep or *'tenshu'*.

Teppan. (J.). Iron plate, used in the protection of gates (*'mon'*) and gate furniture like gate bars (*'kannuki'*) from attack.

See 'hagasane'.

Teppôsama. (J.). A musket port set into the plastered wall or tower of a Japanese castle. In section a port was funnel shaped, the larger part on the inside gave a greater field of fire to the defenders, while the narrower aperture on the exterior provided protection to the defenders from incoming fire. The *'teppôsama'* came in a number of shapes; square, rectangular, triangular and circular. The ports of the plastered walls (*'dobei'*) on the *'ishigaki'* were left open while those of a wall tower were often provided with shutters, since the towers were put to other uses other than defence (e.g. storehouses or residences). Also known as '*juugan*'.

See '*hako hazama*', 'ishisama', 'sama', '*shinogi hazama*', '*taihosama*', '*yasama*', '*yazama*', '*zama*'.

Terpen. (D.). Raised mound, precursor to motte, low countries.

See '*kernmotte*', '*motte*'.

Terrace. Any type of earthworks which were built up above the normal ground level.

See '*artillery terrace*'.

Terrasse. (Fr. *terrace*). **See '*terrace*'.**

Terrapieno. (It. *embankment*). To fill with earth.

Terreplain.

(1) The open country surrounding a field work.

(2) The rear talus of a rampart.

(3) The gun position on the top of a rampart located behind a parapet.

Testudo. (L. *tortoise*). A formation used by Roman soldiers with raise their shields over their heads, with the ranks of the front and sides with their shields down, screening the soldiers from missile attack during siege operations.

See '*tortoise*'.

Tête de pont. (Fr. *head of the bridge*). A fortification on the vulnerable end of a bridge, usually constructed as a '*redan*' . Also known as a '*bridge head*'.

Tetrahedron. A French obstacle, made from timber, metal or concrete in the 3-D shaped pyramid, used in the 1930's CE on hard ground, and could be fitted with mines. Also known as a '*porcupine*' or '*triangular pyramid*'.

See 'beach obstacles'.

Tett Turret. A type of pill box for one or two man revolving turret, formed from a standard 4 foot (1.2 metres) diameter concrete pipe sunk into ground for airfield defence.

Texas barrier. See '*bremer wall*'.

'*T*' Head. A '*T*' shape sap (or '*short parallel*') forward of the front trench, used to break up the line of an enemy attack. Sometimes dug between the '*reserve line*' and the '*support trench*'. Also referred to as an '*advanced line*'.

See '*D*' *shaped flanker*', '*flanking tower*'.

Theodosian (Land) Walls. The walls raised by Theodosius II (404-405 CE) to protect the landward side of the city of Constantinople

(Istanbul, Turkey) superseding the '*Constantinian Walls*', running from the Sea of Mamara (the Marble Tower) to the waters of the Golden Horn (Palace of Blachernai), a distance of 5.7 km (3.5 miles), consisting of multiple lines of defence. The Theodosian walls consist of; the main inner wall ('*mega teichos*', great wall) with towers and gates along its length, the lower outer wall ('*exo teichos*' or '*micron teichos*', small wall) separated from the main wall by a terrace ('*peribolos*') and supplied with smaller towers and gates. Between the outer wall and the moat ('*souda*') was the outer terrace ('*parateichion*'), with a low wall topped the inner edge of the moat with battlements, including arrow slits in the merlons. The Theodosian land walls were finally breached by the Ottoman Turks in 1453 CE using cannons.

See 'Constantian Wall', 'defence in depth', 'Herclian Wall'.

Three-quarter round tower. A cylindrical flanking tower projecting only three quarters of the way from the body of a castle. The rear of the flanking tower was either circular or straight.

Throat. The constricted part of a loop hole or embrasure, which can be at the front, middle or back, acts as a pivot point when taking aim.

See *'cheek', 'embrasure', 'floor', 'loop hole'*.

Tian chuang. (Ch. *sky window*). Larger than regular murder holes, and built behind gate doors, which would allow workers to fill the space behind the gate with earth to prevent the flood waters breaching the gate. Also known as a '*tian jing*'

Tian jing. (Ch. *sky well*). A loophole opened at the base of a parapet between two merlons allowing defenders to drop stones and other substances on the enemy below.

Tian jing, Dou yan and Xuan yan, Great Wall of China.

See 'dou yan', 'xuan yan'.

Timber laced rampart. A rampart which was reinforced by filling the spaces between the upright logs with horizontal ones, and the gaps were faced with dry stone walls, adding to the rampart's strength.

See *'crib work'*, *'rampart'*.

Tituli. (L.). Additional ditch dug forward of a Roman fort gateway.

Titulum. (L.). A short length of ditch and rampart set in front of a gate and parallel to the ramparts of a Roman fort.

Toblerone line. *See 'promenthouse fortified line.*

Tobruk. A defensive position constructed by the Italian army to defend Tobruk in Libya, similar to a *'fox hole'*. Later the Germans fortified such positions with concrete and tank turrets. (i.e. Omaha beach (France), or Saint Aubin's Fort (Jersey)).

See *'ringstände'*.

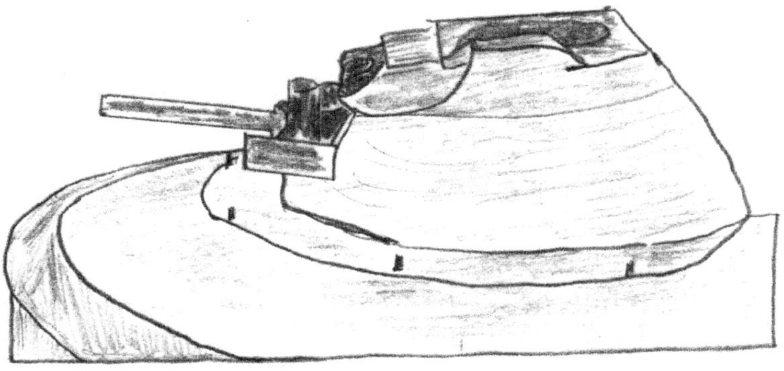

Tojou. (J.).

(1) A separate fortified enclosure outside the main castle but connected to it. Also known as *'sotoguruwa'*, *'sotojiro'*.

(2) A fortified site separate to the main castle, a branch castle (*'shijou'*).

Tomoekoguchi. (J.). A spiral shaped castle entrance (Edo period), consisting of a series of gates connected by walls in a spiral pattern.

See *'koguchi'*, *'umadashi'*.

Tonashimon. (J.). A gateway with no gates, built like a *'koriamon'* but without gates. Supposedly the lack of gates would entice an enemy force to rush in and be trapped before another larger and more fortified gate.

See *'koriamon'*, *'mon'*.

Topfmine. A German anti-tank mine of WWII, circular in shape, made from compressed wood pulp, cardboard and tar, with glass plugs and components, and so was not detectable by Allied forced mine detectors.

See '*mine*'.

Torenfort. (Du. *tower fort*). Dutch artillery tower three stories high from the early 19th century CE, cylindrical in shape, included garrison housing and stores, numerous artillery embrasures and surrounded by a wet moat and low earthen walls.

Torenfort in Muiden, Netherlands.

Torgraben. (G.). Gate ditch.

Toride. (J.).

(1) A suffix used in place of '*jō*', often refers to satellite fortifications.

(2) A small fort built near the main castle, usually defended by a palisade or stockade.

See '*jo*', '*palisade*', '*stockade*'.

Tormenta. (L. *artillery*). Any sort of artillery used to launch rocks, arrows and other projectiles at an enemy.

Torpedo launcher. A land based torpedo launcher placed in specific bunkers, used as part of a coastal defence system.

Torre de las armas. (Sp. *weapons tower*). A Spanish castle's gatehouse.

See '*gatehouse*'.

Torre de homenaje. (Sp. tower of homage). A Spanish keep, provided with a single entrance on the second floor for reasons of security. The parapet was provided with pointed merlons.

The *Castle of La Mota in the town of Medina del Campo, province of Valladolid, Spain.*

Torre de managem. (*Portuguese management tower*). A keep.

Torre massimiliane. (It. *Maximilian tower*). ***See 'Maximilian tower'.***

Torrione. (It. *tower*). An Italian Renaissance transitional gun tower, in a circular or '*U*' shape, provided with an apron, casemates, thick walls, and some with old style decoration of crenulations and machicolations.

See 'gun towers'.

Torri saracinesche. Italian towers built in coastal areas to protect from the attack of the Saracens.

See '*vigla*'.

Tortoise. A small mobile siege tower with a triangular roof, provided with a ram.

See '*testudo*'.

Toter raum. (G. *dead space*). Dead ground.

Totten embrasure. An American type of embrasure, reinforced with iron, provided with smaller openings and wider angle of traverse than most European embrasures of the 19th century CE, usually provided with '*Totten shutters*'. Developed by J. G. Totten, appointed Chief Engineer of the United States Army in 1838 CE.

See '*embrasure*', '*totten shutters*'.

Totten shutters. Iron doors on an embrasure which were balanced to swing freely, the iron shutters would be forced open by the gasses expelled from a cannon firing, and then rebound shut immediately afterwards, shielding the gunners from incoming fire, invented in 1844 CE.

See 'crenel shutter', *'embrasure'*, *'totten embrasure'*, *'turning post'*.

Tour-de-ronde. (Fr. *around the tower*). A gallery running around a tower. An example is the Keep of Provins, France.

Tourell galopin. A French armoured disappearing turret dating from 1907 CE, for an extant turret see Douaumont Fort, Verdun, France.

Tour maitresse. (Fr. *master tower*). ***See 'donjon', 'keep'.***

Tour-Modèle. (Fr. *model tower*). A type of French coastal tower constructed under Napoleon, along the Atlantic coast and Mediterranean coastline, coming in five types, containing powder magazines, store rooms and lodgings for the troops, dating to 1811 CE, by 1814 CE only twelve towers were completed before the project was abandoned. The towers can be compared to the British '*Martello towers*' of a similar date. Also known as an '*empire tower*', or '*Napoleon Tower*'.

Tourelle. (Fr.). A turret projecting from a tower, acting as either a watch tower if provided with machicolations for vertical defence, or just for decoration.

See 'bartisan', 'turret'.

Touret. See *'turret'*.

Tower. (L. *turris*, tower). A tall structure constructed of either timber or stone and used for the purpose of defence, which varied in design depending on the defensive system used, the available materials and skilled labour. The tower was employed singularly or as a part of a fortification, especially at the angles and the entrances; the gateway was commonly flanked by towers provided with numerous forms of defence, such as arrow slits, parapets and machicolations.

See *'flanking tower'*, *'pele'*.

Tower hollow. A rounding made of two brisures so they can join a curtain wall to an orillon, small cannon could be hidden here.

See *'brisure'*, *'orillon'*.

Tower house.

(1)The residential defensible tower of the 14th to the 16th century CE.

See *'z-plan tower house'*.

(2) A fortification which was developed in the Scottish/English border region, which consisted of a simple, strong rectangular tower. They

were not intended to meet the threat of a major siege but were more suited to the border warfare which was waged in the region.

See *'pele'*.

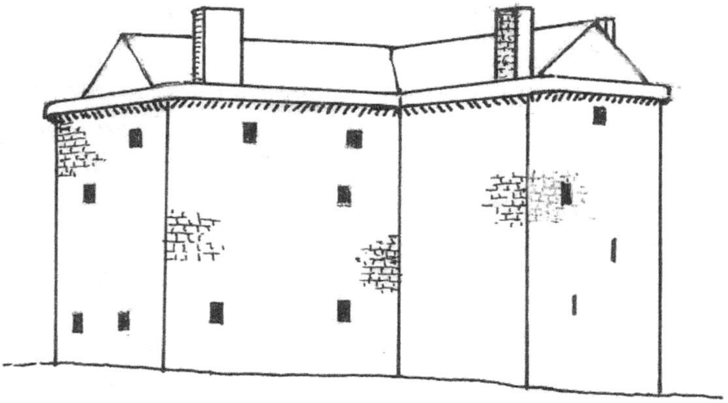

'L-plan", a tower house built on a 'L' shaped plan.

Tower keep. A term which has been used to refer to a number of different types of keeps from different periods.

See *'polygonal keep'*, *'square keep'*, *'transitional keep'*.

Town (fortified). See *'burg'*, *'burgus'*, *'burh'*.

Trace.

(1) The ground plan of a fortification.

(2) A type of fortification.

See *'bastion trace'*, *'polygonal trace'*, *'tenaille trace'*.

Trace Italienne (Fr. *Italian outline*). A bastion system often combined with ravelins and other outworks, developed in Italy in the mid 15th century CE as a response to use of cannon in siege warfare which made earlier medieval castles obsolete.

See *'bastion fort'*. *'bastion system'*, *'trace'*.

Tracé pseudo-bastionné. (Fr. *pseudo-bastioned trace*). A trace consisting of a bastion front with small flanks with entrance and rear ditch can be defended.

See 'bastion trace'.

Traditore batteries. Batteries which provided fire only to the flanks and rear.

See *'batteries'*.

Trajan's wall. See *'Limes germanicus'*.

Tranchee. (Fr. *trench*).

Transitional keep. The polygonal or round keep of the 12th century CE, which was designed to overcome the problems associated with square and rectangular keep. The purpose of the circular and polygonal forms employed was to avoid giving an enemy any cover around the base of the keep, known as dead ground, from where the enemy could launch attacks as well as undermine or batter the walls. The rounded and polygonal edges also offer a greater resistance to battering.

See *'keep'*, *'polygonal keep'*.

Trap holes. Pits dug in the front of a field fortification with a sharp wooden stake at the bottom, set out in a checkerboard pattern, effective against infantry and cavalry.

See *'lilia'*, *'trous-de-loup'*.

Traverse.

(1) A mound or earth situated at intervals along a work such as a covered way, the traverse were aligned at right angles to the work and prevented it from being swept by flanking fire should part of the work be taken by the enemy, or observed from higher ground by the enemy.

(2) A work similar to a caponier, consisting of a gun looped passageway which traversed a dry ditch of a fortification, and was used to sweep the ditch of the enemy as they tried to cross the ditch.

See *'caponier'*.

(3) A parapet crossing the covert-way, situated opposite the salient angle of the works and near to the places of arms, used to prevent enfilade fire.

(4) Traverses were also built in a caponier, and were known as *'tambours'*.

Traverse shelter. Protective shelters made of stone and earth on the parapet of the traverse of 19th century CE polygonal forts, provided protection from enfilade fire and came in handy for ammunition stores.

***See 'traverse'*.**

Traversing platform. An emplacement, circular or semi-circular in shape, on which a gun could be mounted *'en barbette'* on a movable rail.

Tread. The part of a banquette where a soldier stood to fire over the parapet.

See *'banquette'*, *'fire step'*.

Trebuchet. A type of siege artillery that came in two types.

(1) *'Counterweight'*. A counterweight trebuchet was supported on two triangular side supports, with a pivot bar between them which housed the pivot arm from which hung a box of stones acting as a counterweight on the short end of the pivot. A sling was attached to the

long end of the pivot arm. The short end of the pivot arm is pulled down using a winch, sometimes powered by man powered wheels. The sling was loaded under the pivot arm and then released, producing a high trajectory over castle walls. Anything could be loaded into the sling; stones, dead animals, human heads.
(2) '*Traction*'. A traction trebuchet with a single support beam or two triangular side supports. The power was supplied by a team of soldiers pulling down on ropes attached to the short end of the pivot arm. The traction trebuchet could be used in attack and defence.
Trench. A ditch dug in the ground; the earth from the trench was thrown up to form a parapet, used as a defensive or offensive position.

See '*sap*'.

Trench block. A movable obstacle made of wooden frame covered in barbed wire, used to block trenches (usually in the '*communication trenches*') to hinder the movement of enemy troops when invading the opponent's trenches.
Trench cavaliers. A series of high parapets built from gabions which are pushed forward to top the covered way of a besieged fortress.

See '*covered way*', '*gabion*', '*parapet*'.

Trench digging machine. A machine used to dig trenches used for trench warfare, Prototypes were developed but not deployed in WW1, including the British '*Cultivator no. 6*' and the German '*A7V-Geländewagen*'. Russian developed trench diggers from the truck based bucket digger '*BTM-3*' to the tracked traverse digger '*MDk-3*' (which has been in use during the Ukrainian / Russian war in early 21st century). The Ukrainians developed their own armoured trench machine, the '*PZM-3*'.
Trench fascines. Fascines used for crowning a line of '*gabions*' in sap or trench, made 1.2m to 1.8 m (4 to 6 feet) long.

See '*fascines*'.

Trench parapet. The front wall of a trench, closest to the enemy, which could be strengthened with wooden planks Used to protect soldiers standing on the '*fire step*'.
Trench system. The trench systems erected during WWI on the Western Front were about 764 kilometres (475 miles) long, stretching from the North sea coast of Belgium southward through France to the Swiss Alps, although not in a continuous line.

See 'linear fortification'.

Trench Trace. The form of trenches used in WWI, consisting of a range of elements.

Trench warfare. Warfare where both sides dig in and defend territory with a system of opposing trenches dug into the ground incorporating barbed wire, machine guns and artillery, separated by a '*no man's land*'. The use of trenches and other obstacles to defend a front line (WW I and Ukraine vs Russia 2022-2024 CE).

See '*trench system*'.

Tribok. (G. *trebuchet*). A trebuchet from 13th century CE Germany, with triangular supports on either side of the tossing beam.

Trickery. to gain access by means of ruse or treason (i.e. Saracen capture of Krak des Chevaliers, Syria).

See '*dirty tricks*'.

Trip wire. A wire set to catch someone's foot or lower leg and set off a trap or alarm.

Trous-de-loup. (Fr. *wolf holes*). An infantry obstacle in front of the glacis of a fortification, consisting of a hole in the ground measuring about two metres deep and in diameter, with a sharpened wooden stake at the bottom.

See '*lilia*', 'trap holes'.

Trutzburg. (G.). A strongly built castle able to resist attacks. Also known as a '*gegenburg*'.

Tschechenigel. (G. *Czech hedgehog*). ***See* '*czech hedgehog*'.**

Tsubosama. (J.). Loopholes through castle walls made from bamboo for firing guns or loosing arrows through, which are filled with dirt during peace but the dirt is removed during times of war.

See '*sama*'.

Tsuchido. (J.). The plastered window shutters which were used to cover the loopholes of a Japanese castle, which provided protection to the defenders within from incoming fire, the plastering reduced the risk of fire.

Tsuchisama. (J.). Loopholes made through earthen embankments or walls of a castle.

See '*sama*'.

Tsujiumadashi. (J.). A barrier built at a ninety degree angle at the intersection of two entrances to a castle.

See '*umadashi*'.

Tsukejiro. (J.).

(1) A fortified camp with one or two enclosures, embankments and moat. Also known as '*mukaijiro*' (*facing castle*) or '*tai no shiro*' (*confronting castle*).

(2) A branch castle ('*shijou*').

See *'counter castle'*, *'malvoisin'*, *'shiro'*, *'tai no shiro'*.

Tsume no shiro. (J.).

(1) A castle at the rear of a main castle, which can be retreated to if the main castle falls to an attack.

(2) The central ward (*'honmaru'*) as opposed to the other wards.

(3) The main castle (*'honjou'*).

See *'honjou'*, *'honmaru'*, *'jiro'*, *'shiro'*.

Tsunagi no shiro. (J.).

(1) A branch castle built on a strategic position between two main castles, used to maintain communications. Also used by soldiers moving between the two main castles.

(2) The central ward (*'honmaru'*) as opposed to the other wards.

See *'honjou'*, *'honmaru'*, *'jiro'*, *'shijou'*, *'shiro'*, *'tsume no shiro'*.

Tsuribei. (J.). A false wall hung over an earthen embankment or stone wall of a castle, which could be released to fall on the attacking soldiers.

See *'bei'*.

Tsurugibe. (J. *sword walls*). Rows of sharp metal spear-like protrusions for wall tops intended as anti-personnel devices, like *'barbed wire'*. Examples can be found at Nagoya, Kochi and Kumamoto castles.

Tsuzuki yagura. (J.). A wall tower of a Japanese castle which was usually only a single storey in height, which functioned as a connecting tower or gallery running along the top of the ishigaki, from which the defenders protected the walls.

See *'yagura'*.

Tuijibei. (J.). A variation of *'dobei'* constructed from a compacted mixture of clay and sand, about 3 metres (9 foot 10 inches) tall and a metre (3 foot 3 inches) thick, with a wooden front and tiled roof.

See *'dobei'*.

Tulou. (Ch.). A large circular or rectangular fortified structure enclosed by walls formed by compacted earth, with stone, bamboo, wood, with outer walls 1.8m (6 foot) high, and towers (built of the same materials) three to five stories high, housing up to eight hundred people, along with storehouses, halls, wells and accommodation. The main gate was closed by wooden doors reinforced with iron. The upper levels were provided with gun loops. Mainly built between the 12th and 20th century CE in Fujian province, China.

Tunnels. Underground access to different parts of a fortification providing communication and protection. Examples: Krak de cheverals (Syria), Maginot line (France), Iwo jima (Japan).

Turf rampart. A rampart made of uniform bricks of turf stacked to form a vertical wall which was invulnerable to both fire and battering ram.

See ***'rampart'***.

Turm. (G. *tower*).

Turning bridge. See *'bascule bridge'*.

Turning post. A timber cylinder located in an embrasure with a vertical oblong hole in its centre for firing a gun through, the cylinder was then rotated to block up the embrasure from incoming fire. Dating from the 15th century CE, see the Laufer gate, Nuremberg. Germany.

Turret, Turette, Touret. (L. *turris*, tower).

(1) A small tower or bartizan, which was often placed at the angles of a castle, to increase the flanking ability, some only serving as corner buttresses.

See ***'bartisan'. 'corner buttress turret', échauguette', 'flanking turret', 'kapishirsha', 'machicolated turret', 'machicoulis', 'staircase turret', 'tourelle', 'watch turret', 'yagura'.***

(2) An armoured turret of a defensive line (i.e. Maginot Line), consisting of a thick steel cap with steel walls, resting on a pivoting tube, which could house machine gun(s), mortars, cannon and mixed arms.

See 'Allan Williams turret', 'cloche', 'Coles turret', 'eclipsing turret', 'Galopin turret', 'Gruson turret', 'Habicht turret', 'Maginot Line', 'Mougin turret', 'panzerturm', 'round gun turret', 'sechsschartenturm', 'tett turret', 'tobruk'.

Turret cap. A metal covering for an armoured turret.

Turreted. Provided with turrets.

Tutulus. (L. *safeguard*). A separated length of earthen bank and ditch screening the opening of the gate of a Roman camp, to prevent direct approach.

See 'clavicula'.

Tvrz. (Czech). A citadel, fort, fortress or stronghold.

T-wall. See *'bremer wall'*.

Twydall profile. A slope that runs from the crest of the fortification to about 15 to 20 feet (4.6 – 6.1 metres) below the natural ground level, with a steel palisade and barbed wire entanglements at the base, sometimes used in place of a ditch, used in British polygonal forts dating to the end of the 19th century CE. Examples are the redoubts built near Chatham in 1886 CE. The name '*Twydale*' comes from the name of the village in Kent where the first of the forts using this profile were used.

See *'dacoit fence'*, *'profile'*.

U.

Ubaldini Migliorino. Italian fortification engineer, 16th century CE.

U-Boot bunkers, U-boat pens. (G. *Unterseeboot*, Underwater boat / submarine). The Germans defended harbours for U-boats on the French coast. Brest (France) had 5 wet and 5 dry (narrower) pens, with a roof 3.5 metre (11 foot) of concrete, then a layer of steel and more concrete 5.5 metre (10 foot) thick, with anti-aircraft guns on the roof.

See *'submarine pen'*.

Uchibori. (J.).

(1) The inner moat of a castle that has two or more moats.

(2) A secondary moat dug on the inside of the stone wall of a mountain castle.

See *'bori'*, 'hakobori', *'horikiri'*, 'karabori', *'kenukibori'*, *'moat'* *'shojibori'*, *'unebori'*, *'unejotatebori'*, *'yagenbori'*.

Uchiguruwa. (J.). An inner compound or bailey of a castle, usually in the central compound, this may include other compounds such as the *'honmaru'*, *'ninomaru'*, and *'sannomaru'*. Also known as *'naikaku'* and *'naijou'*.

See *'honmaru'*, *'kuruwa'*, *'ninomaru'*, *'sannomaru'*.

Uchikomihagi. (J.). Ishigaki where the stones are worked and the gaps between the large stones are filled with smaller stones, this type of wall is stronger than *'nozurazumi'*, so is suitable for higher walls.

See *'ishigaki'*, *'nobori ishigaki'*, 'kikkozumi', *'kirikomihagi'*, *'nozurazumi'*, *'nunozumi'*, *'ranzumi'* 'tamaishizumi',*'tanizumi'*.

Uchimasugata. (J. *inner square box*). A walled compound, square in shape, with a small gate, leading to the square compound, and then soldiers need to make a right angled turn to enter the second gate. The purpose is to slow down an attack by enemy troops in the killing ground of the compound.

See *'masugata'*, *'sotomasugata'*.

Uchi umadashi. (J. *inner horse exit*). A defensive barrier at the entrance of an inner compound of a castle with many compounds, usually built at the entrances to the second and third compounds.

See *'nino maru'*, 'sanno maru', *'umadashi'*.

Ukeks. An 11th century CE flanking tower of a Turkish *'baliq'* or small fortified town.

See *'baliq'*.

Umadashi. (J.). A barbican or barrier providing extra defence at the entrance to a Japanese castle. Earthen bridges (*'dobashi'*) were provided forward and aft of the '*umadashi*' and could not be easily destroyed. For mountain castles the 'umadashi' were protected by dry moats ('*yamajiro*') while flat land castles ('*hirajiro*') used wet moats.

See ***'azuchi umadashi', 'barbican'***, 'dobashi', ***'doi', 'hori', 'kaku umadashi', 'koguchi', 'maru umadashi', 'umadashi'.***

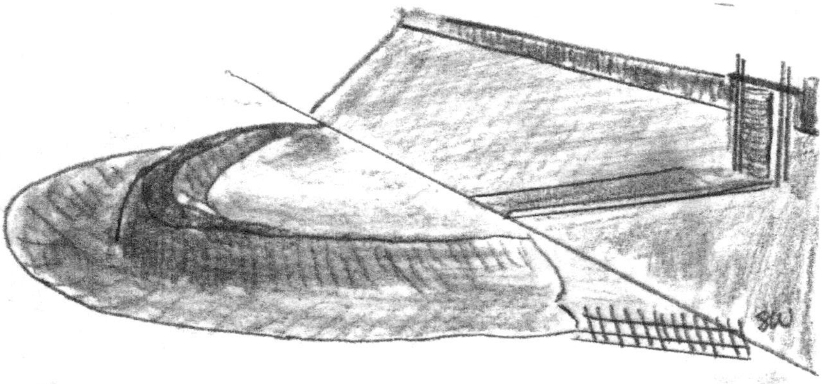

Umadashi koguchi. (J.). The main gate of a Japanese castle provided with the extra protection of an '*umadashi*' or barbican, using earthen embankments and moats, along with screens ('*shitoma*') and staged embankments ('*kuichigai doi*') often sited inside the gates.

See 'doi', *'koguchi', 'masugata', 'shitoma'*, 'umadashi'.

Umijiro. (J.). A castle on the edge of the sea.

See *'jiro'*.

Underground hangar. A type of military aircraft hangar built underground or into a mountain, used to store fighter planes, may include; fuel and weapons storage, maintenance, accommodation and communications. An example is used by the Royal Swedish Air force.

Unebori (J.). A trench with ridges dug across them of a Japanese castle.

See 'bori', 'hakobori', 'horikiri', 'kenukibori', 'shojibori', 'tatebori', 'unejotatebori', 'yagenbori'.

Unejotatebori. (J.). Multiple parallel '*tatebori*' (trenches) across one slope on a mountaintop castle.

See 'bori', 'hakobori', 'horikiri', 'kenukibori', 'shojibori', 'tatebori', 'unebori', 'yagenbori'.

Usine. (Fr. *factory*). A power plant of the '*Maginot Line*' which produced and converted alternating current (AC).

See 'Maginot Line'.

Usume mon. (J. *embedded gate*). A gateway which was used to defend the gate of a Japanese castle, the gateway was a simple opening made in the *'ishigaki'* and defended in much the same way as other such gateways.

See *'mon'*.

Unterschlüpfe. (G. *shelters*). Dug out.

Unterständ. (G. *shelter*). A generic term for a German WWII concrete bunker or shelter.

Unterständ für Mannschaften. (G. *shelter for teams*). Personnel bunker or shelter.

Unterständ für Munitions. (G. *shelter for ammunition*). Ammunition bunker.

Unterständ für Waffen. (G. *shelter for weapons*). Weapons bunker.

Utum. (Ar.). An Arabian tower or refuge at an oasis.

Uzumimon. (J. *buried gate*).

(1) A gate which goes through the stone walls of a Japanese castle, connecting areas on two different levels, and the area behind the gate could be filled in if the castle came under attack, effectively blocking that entrance.

(2) A back gate in a stone wall or a hard packed mud wall (*'tsuiji'*) of a Japanese castle.

See *'mon'*.

V.

Vainakh. (Chechen). Residential or military stone tower, square in cross section (ranging from 6 to 12 metres (19 to 39 feet) in width at base, and 10 to 24 metres (32 to 78 feet tall, as the wall raised the width decreased. Earliest towers date from 1st century CE, with increased construction in the 12th to the 13th century CE but continued on into the 15th to the 17th century CE. Also known as a *'nakh tower'*.

See '*casatorri*' in Italy, '*Svan*' in Georgia, '*vepshi*' in Ingushetia, Russia.

Valli, Vallus. (L.). A stake carried by a legionarius to fortify a rampart on the march. The sharpened stakes which were lashed together and rammed into the top of the rampart surrounding a Roman encampment.

See '*sudis*'.

Vallo Alpino. See '*Alpine wall*'.

Vallum. (L. *wall*).

(1) A single stake of the palisade topping the rampart surrounding a Roman encampment.

(2) The palisaded rampart or the rampart itself surrounding a Roman fortification.

Vamure.

(1) A parapet built on the outside of a rampart.

(2) An advanced wall or earthwork built in front of a main fortification.

Vana-durga. (Sanskrit). An Indian forest fort that came in two variations, see *'khanjana durga'*, *'stambha durga'*. Also known as *'vrikshya durga'*.

***See* '*durga*'.**

Vapra. (Sanskrit). A rampart of a fort made of spoil from digging the moat.

See '*rampart*'.

Vauban, Sebastian le Prestre de Vauban, Marshall of France (1633 – 1707 CE), designer and besieger of fortifications.

Vauban's three systems. First system, set up a system of bastions which mutually protected each other, each face was protected by a neighbouring bastion.

Second system, the second system increased the number of walls (called the *'combat wall'* and the *'safety wall'*) and added bastioned towers to the ramparts, all of which provided a defensive belt around

the strong hold, increasing the fire power brought to bear on the attacking enemy.
Third system, the third system of Vauban consisted of improving the previous systems, to further increase the resistance of the stronghold. See Neuf-Brisacha, a town in the commune of the department of Haut-Rhin in the French region of Alsace.
Vawnmewre. An earthen rampart used as a defence against artillery.
Venetian battlement. A battlement provided with stepped merlons.

See '*Irish battlement*'.

Ventilation shaft. The vents in the casemate ceilings for the removal of toxic fumes from cannon or musketry fire.

See '*air hole*'.

Vepshi. (Ingush). Tower, square cross section tapering in width upwards up to four or five stories, with small windows on lower levels and larger window as you rise, 15 to 25 metres (49 to 82 feet) tall, from the Republic of Ingushetia, Russia. Used as fortified homes and watch towers, dating from between the 13th to the 17th century CE. Similar to '*svaneti*' in Georgia and '*casatorri*' in Italy.
See '*casatorri*' in Italy, '*vainakh*' *in* Chechnya, '*svan tower*' in Georgia.

Verbindungsgräben. (G.). Fire trenches parallel to the front line. Also known as '*kampfgräben*'.

See '*trench*'.

Verstärkt feldmässig, (Vf). (G. *reinforced field*). A building built to the strength of a reinforced field fortification.
Vesting Holland. (Du. *Fortress Holland*). The Dutch national réduit consisted of fortifications and flooding built between the 17th century

and 1940 CE, from Naarden to Moerdijk defending the Randstad (85km, 53 miles).

See 'Dutch water line','Grebbe line'.

Via decumana. (L. *ten-year-old road*). The road leading to the rear gate (*porta decumana*) from a '*T*' intersection with the '*via quintana*'.

Via praetoria. (L. praetorian road). The road leading from the main gate (*porta praetoria*) to the headquarters (*principia*) at the intersection with the '*via principalis*'.

Via principalis. (L. *main road*). The pathway which followed the two intersecting lines bisecting a Roman encampment, which were approximately thirty metres wide.

Via quintana. (L. *fifth road*). The road that runs parallel to the '*via principalis*', forming a '*T*' intersection with the '*via decumana*'.

Via sagularis. (L. *grass road*). The street that runs around the interior of the '*intervallum*' of a Roman fort.

Via vicinariae. (L. *neighbourhood road*). The ancillary streets between the tents or buildings of a Roman fort.

Vickers Medium Machine Gun (MMG) emplacement. A British WWII emplacement, square in plan with front corners chamfered with a blast wall in front of the entrance in the rear. The other three embrasures in the other walls could be defended by rifles or light machine guns. The Vickers Machine Gun emplacements were often sited in pairs. The Vickers Machine Gun emplacements were uncommon with seventy five recorded as being extant.

See 'British hardened field defences', 'FW3', 'pillbox'.

Victoria lines. A defence line along the width of Malta, 12 km (7 ½ mile) long, built by the British in the late 19th century CE, consisting of batteries, entrenchments, forts, Infantry lines, search lights and stop walls. Also known as the '*north west front*'.

See *'Corradino lines,' 'Cottonera lines', 'Floriana Lines', 'Santa Margherita Lines'.*

Vigilarium. (L. *watchman*). A Roman watch tower on a '*limes*', which was surrounded by a wall and ditch. Also known as a '*specula*' or '*burgus*'.

See *'burgi', 'limes', 'specula'.*

Vigels. (L. *vigla*, watch, Gr. βίγλα). The network of watchtowers of Chios (the fifth largest Greek island, in the northern Aegean Sea), built by the Genoese between 1346 and 1566 CE and still in use until the 18th century, sited close to each other to aid in message transmission if pirates were sighted. Currently, 35 '*vigels*' can be identified. The structure of a '*vigel*' is cylindrical, with an average diameter of 7.50 metres (24 feet) and a total height of about 12 meters (39 feet), topped by a parapet with square merlons, access to the only room was a door at 8m (26 feet). Also known as a '*pyryktoria*' in Greek or '*kaminoviglia*' by the Byzantines.

Viking ring forts. See '*ring forts (Viking)*'.

Village Perche. (Fr. *perched village*). A French fortified village from the 9th to 10th century, set on isolated hill tops surrounded by earthen rampart.

Vineae, (L. *vineyards*). Movable sheds used by besiegers as overhead protection when advancing to the walls and when operating siege equipment or undermining the castle's wall.

Visvasghati. (Sanskrit). A wooden beam set above an entrance of a fort when released would fall upon attackers.

Vitrified Hill Fort. A hill fort provided with vitrified walls. Examples can be found in Scotland and Sweden.

See '*hill fort*'.

Voldsteder. Danish mottes of the 12th century CE.

See '*motte*'.

Vorberg. (G.). Bailey.

Vorpanzer. (G. *pre-armour*). Forward or glacis armour.

See '*avant-cuirasse*'.

W.

Wagenburg, Wagon fort. (Czech. *vozova hradba*, Wagon wall). A field work made by chaining wagons together in various shapes (circular or square), with pavises in the gaps between wagons, defended with hand to hand weapons, handguns and cannons, used by the Hussites (Followers of Jan Hus, a Bohemian religious reformer).

See *'carrago'*, *'gulaigorod'*, *'laager'*.

Wagner line. A defence line built in eastern Ukraine, up to the Russian border (200 km, 120 miles) by the Wagner paramilitary group in 2022 CE, consisting of two parallel ramparts, a trench and firing stations.

See 'Surovikin line'.

Wajõ. Wajou. (J.), **Waeseong** (Korean). A Japanese castle built in Korea in the late 16th century CE during the '*Bunroku-Keichou*' war, mainly along the coast or with access to the sea. The '*wajou*' consisted of three compounds ('*honmaru*', '*ninomaru*' and '*sannomaru*'), *with ramparts, moats, walls, gates and towers.*

See *'honmaru'*, *'ninomaru'*, *'sannomaru'*.

Wakikoguchi. (J.). A small side entrance to a castle, used by soldiers to counter attack an assailing enemy, as opposed to the main gate ('*ootemon*') and the rear gate ('*karametemon*').

See *'koguchi'*, *'postern'*.

Wall head. The apex of a curtain wall or rampart.

Wall head defences. Defences which were used to defend a fortification from the top of the walls.

See *'brattice'*, *'crenel'*, *'hoard'*, *'machicolation'*, *'merlon'*, *'parapet'* and *'turret'*.

Wall tower. A tower projecting from the curtain wall of a castle, which provided both neighbouring wall towers and curtain wall with flanking fire.

See *'curtain tower'*, *'flanking tower'*, *'mural tower'*.

Wall walk.

(1) A walkway along the top of an embankment of a rampart. Also known as a *'bailey walk'*.

(2) A walkway situated behind a parapet.

See *'alatorium'*, *'bailey walk'*.

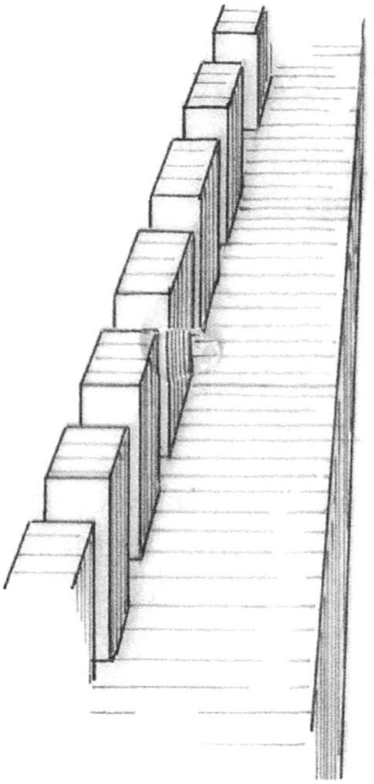

Walls of Benin. ***See benin moat'.***

Walls of Hattusa. Capital of the Hittite Empire in the late Bronze Age, toward the end of the 3rd Millennium BCE. Hattusa's ruins lie near modern Boğazkale, Central Anatolia Turkey. Hattusa had 6 km (3.7 miles) of surrounding walls, with inner and outer walls around 3 m (9.8 feet) thick and with 2 m (6 feet 7 inches) of space between them, making 8 m (26 feet) of total thickness.

Ward. (O.E. *weard*, guard). A courtyard of a castle, or the space between the lines of bastioned curtain walls.

See *'bailey'*, *'lists'*.

Wasserburg. (G. *wasser*, water; *burg*, castle). A German term for a castle built on a bank or island of a river, making use of the water way in their defence. Their main purpose was to extract tolls from the trade being shipped on the water way.

See *'jala durga'*. *'mizuki'*, *'water castle'*.

Watari yagura. (J.). A connecting tower or crossing wall tower of a Japanese castle, which ran along the *'ishigaki'* and was used by the defenders to protect the walls against assault.

See *'yagura'*.

Watari yagura. (J.).

(1) A connecting tower or crossing wall tower of a Japanese castle, only a single story in height, which ran along the *'ishigaki'* and was used by the defenders to protect the walls against assault. Provided with gun ports (*'teppousama'*).

(2) A long connecting gallery or corridor, built on top of a stone wall (*'ishigaki'*) Also known as a *'tamon'* or a *'nagaya'*.

See *'watari yagura'* *'yagura'*.

(3) A long walkway made using wood planks along the *'parapet'* used by archers to loose arrows.

Watari yagura mon. (J.). The largest kind of Japanese castle gateway, which consisted two storey timber frame structure, where the lower storey incorporated the entrance, and the upper storey formed the connecting tower (from where the gateway derives its name).The style of the connecting tower is typical of a yagura but the lower storey's timber is left unplastered, and the actual gate and surrounding timbers were often covered with strips or sheets of iron which improved their protection against battering and firing.

See *'yagura mon'*.

Watch turret. A watch tower or *'bartizan'*.

Water battery. A detached work with a group of cannons facing the sea or waterway.

Waterburcht. (Du. moated castle). A Dutch medieval castle surrounded by a wet ditch.

Water castle. A castle which used water defences of natural or artificial origin, or both.

See *'wasserburg'*.

Water defences. Water filled features used for defence.

See *'lines of defence'*, *'moat'*.

Water fascines. Fascines used as cover on marshy ground, made 1.8 to 2.7m long and 46 cm in diameter (6 to 9 feet long and 18 inches in diameter).

See '*fascines*',

Water gate. A defensible gate which connected a castle to open water or a water way, which was used to supply a castle with stores, reinforcements and communications, especially useful in times of siege.

Water gate (Traitors' Gate), Tower of London, UK.

Waterline. The Dutch system of national defence based on the use of inundations and fortifications.

See 'Dutch water line', 'Grebbe line', 'vesting holland'.

Wat's dyke. A ditch and rampart running from the Dee estuary to Oswestry in England, 56 km (35 miles) long, east of and partially to Offa's Dyke, attributed to the predecessor of Offa, of King Wat but could date from the 5th to the early 9th century CE.

See 'Offa's dyke'.

Way of the rounds. A narrow covered way between the rampart and the wall of a fortification.

Weapon storage area (WSA). A well defended and secure ammunition storage. An example would be an army or air force base or the opposite of an isolated site where NATO stored the nuclear weapons during the Cold War. Also known as '*special ammunition storage*' (SAS).

Weapons storage and security system (WS3). A system used to safely store nuclear bombs in the floors of a '*protective aircraft shelters*' (PAS), consisting of electronically controlled vaults at NATO military airfields.

Also known as '*weapon security and survivability system*' (WS3) or '*weapons survivability and security system*'.

Wehrgang. (Gr. *wehr*, defence; *gang*, passage).

(1) A covered gallery provided with machicolations which ran around the perimeter of a castle of the Teutonic Knights, which was accessible from the chapel, the chapter-house, and the dormitories by staircases in the walls, from which the castle was defended.

(2) A battlement.

Wehrturm. (G. *defensive tower*). Defensive or wall tower.

Wellblech understand. (G. *corrugated iron shelter*). A cheap German shelter made from corrugated iron sheets with a layer of concrete and earth.

See '*Anderson hut*', '*Morrison shelter*', '*Stanton air raid shelter.*

Wellington barrier. A barrier consisting of fortified cities in southern Belgium developed in 1815 CE, after the Napoleonic wars, to counter French aggression, supervised by the 1st Duke of Wellington.

Weng cheng. (Ch. *urn fort*). A fortified wall that completely encloses the gatehouse, either rectangular or semicircular in shape, built to the same thickness and height as the main city wall. The gate of the '*weng cheng*' is built out of alignment with the city gate and drawbridge to make the approach more difficult. The '*weng cheng*' could be detached and act as a barbican, thus adding to the depth of defence.

See 'barbican', '*cheng*', '*defence in depth*'.

Werk, werki. (G. *work*). Fortification or heavy bunker.

See '*work*'.

Werkgruppe. (G. *group of works*). A group of bunkers or forts.

West wall. See '*Siegfried line*'.

Wicket. (Fr. *guichet*). A small door in the main gate of a fortification, which could be used without having to open the gate.

See '*guighet*', '*klinkett*', '*postern*', '*sally port*'.

Widerstandsnest (WN). (G. *nest of resistance*). A German WWII defensive position on the '*Atlantic wall*', consisting of armed bunkers and shelters, defended by barbed wire and anti-personal mines overlooking the beaches.

See 'Atlantic wall', 'bunkers','stutzpunkt', 'stutzpunktgruppe'.

Wings. See *'jambs'*.

Wingwall. A palisade built running up the slope of a motte to protect the stairway leading to the main tower on top of the motte.

See 'motte and bailey castle'.

Winkel bunker, Winkeltürme. A type of above ground bunker, conical in shape coming to a point at the top, standing 18 metres (59 feet), designed by Leo Winkel in the 1930s CE to be an air raid shelter for about 100 people. The design relied on the shape to deflect descending bombs, the fact that it was above ground allowing the explosive pressure to dissipate into the surrounding atmosphere. The design did not require as much material, it was cheaper and easier to build, only one of the two hundred built in Germany was destroyed by a bomb. Also known as '*spitzbunker*' or '*winkeltürm*'.

See *'Anderson hut', 'bunker' 'flak towers (Germany)', 'Morrison shelter', 'Stanton air raid shelter'.*

Winter Line. The three German defensive lines across Italy in WWII (*'bernhardt line', 'gustav line', 'hitler-Line)* which was built to prevent the Allied troops advancing up the peninsula, consisting of concrete bunkers, gun emplacements, turreted machine gun emplacements, barbed wire and minefields, which took nine months to overcome.

See 'bernhardt line', 'gustav line', 'hitler-Line.

Wire break. A passage through the wire entanglements used for offensive operations so that the troops wouldn't be hindered by its own wire tangles when on the attack.

Wirkungbereich. (G. area of effect). Field of fire.

See 'schussfeld'.

Wo pu. (Ch. *shack*). ***See 'cheng pu'.***

Works. Forts and fortifications.

See *'breastwork', 'crownwork', 'detached work', 'earthwork', 'forework', 'hornwork', 'low artillery outworks', 'main works', 'outwork', 'salient work'* and *'siege works'*.

Wolfsgrube, Wolf's pit. (G.). A pit with a sharpened wooden spike in the bottom.

See 'lilia', 'trap holes', '*trous-de-loup*'.

Wu xing chi. (Ch. *five star pond*). A special trough built directly over the gate doors with up to five drain holes, which was filled with water to put out fires on the gate doors. If necessary other harmful substances could be poured down on the enemy below.

X.

XPM. See '*expanded metal plate*'.

Xuan ban. (Ch. *hanging plank*). A plank attached to a rail just below the roof to provide protection to the '*zhan peng*'.

Xuan lian. (Ch. *hanging curtain*). A temporary crenel shutter made from a soaked cotton blanket.

See '*pi lian*', '*xuan hu*'.

Xuan hu. (Ch. *hanging window*). A temporary crenel shutter made from timber.

See '*pi lian*', '*xuan lian*'.

Xuan yan. (Ch. *hanging hole*). A loop hole open at the base of a parapet in between two crenels, similar to a '*tian jing*'. Later, a vertical groove below the loop hole was added to better control the foot of the wall.

See 'machicoulis' '*yan*', '*dou yan*', '*tian jing*'.

Y.

Yagenbori. (J.). Shallow *'V'* shaped trenches used to defend Japanese castles, from (*'yangen')* a mortar used to grind herbs).

See 'bori', 'hakobori', 'horikiri', 'kata yagenboi', 'kenukibori', 'shojibori', 'tatebori', 'unebori', 'unejotatebori'.

Yagura. (J. *'ya'*, arrow; *'kur'a*, storehouse).

(1) A watch tower consisting of a wooden frame and no roof which was only raised in time of war.

See *'seirou'.*

(2) A wall tower or a guard tower built atop the *'ishigaki'* which were used by defenders to protect the walls of a Japanese castle. There were four main types and they are as follows. the *'tamon yarura'*, a continuous covered gallery; the 'sumi yagura', a corner tower; the *'tsuzuki yagura'*, a connecting tower; and the *'watari yagura'*, a crossing wall tower. The *'yagura'* not only functioned as a military defence but were also used as storehouses and residences.

See 'sumi yagura' 'tamon yarura', 'tsuzuki yagura', 'watari yagura', 'yagura', 'yaguramon'.

Yaguradai. (J.). A platform for a turret.

See *'yagura'.*

Yagura mon. (J.). A large two storey gatehouse of a Japanese castle. The gateway passed through the first floor, while the second storey housed the guard room and the defensive positions used by defenders to protect the gateway. This type of gatehouse usually constituted a part of a *'masugata mon'*, forming one side of a barbican, adjacent to a gate called a *'kôrai mon'*. There are two types of yaguramon: (1) The first utilises a '*watariyagura*' to span the gateway. (2) The second is a free standing yagura with a gate through the first floor.

See *'kôrai mon', 'masugata mon', 'mon', 'watariyagura', 'yagura', 'yakuimon'.*

Yakatajiro. (J.). A Japanese castle combined the living quarters of a mansion with the defensive aspects of a castle.

See *'jiro'.*

Yakuimon. (J.). A gate where one roof covers the front and rear support pillars. Later replaced by a '*koraimon*', which was more defensible.

See *'kôrai mon', 'masugata mon', 'mon', 'watariyagura', 'yagura'.*

Yamame uchikomizumi. (J.). A type of dry stone wall method used for mountain castles ('*yamajiro*') by piling fan shaped rocks on their sides, leaving the faces unhewn.

See *'ishigaki'.*

Yamajiro. (J. *mountain castle*). A Japanese castle which was usually built on a mountain top or some other elevated position, constructed using walls of rocks and earth to reinforce the natural defences of the site, a moat was optional. This form of castle was the favoured form prior to the latter half of the 16th century CE, where the lessening of civil strife began to show an effect.

See *'hirijiro', 'hirayamajiro'.*
See *'Appendix 2. Shiro - a Japanese Castle?'.*

Yamajiro no koguchi. (J.). The entrance ('*koguchi*') of a Japanese mount castle ('*yamajiro*') was smaller in size than a flatland castle ('*hirojiro*') but did not have a strict format. The difference between the two types of castle was the elevation of the mountain castle and that the sloping approach to the entrance ('*sakakoguchi*') was on angle, thus exposing the attackers to flanking fire from the castle.

See *'koguchi'.*

Yan. (Ch. *hole*). The various types of holes to fire guns and loose arrows at an enemy.

See 'chong yan', 'da jiang jun chong yan' 'dou yan', 'jian yan', 'tian jing', 'xuan yan'.

Yanak. (Sanskrit). A tower on top of a fort provided with leather exterior as protection for archers.

Yang ma qiang. (Ch. *sheep and horse wall*). A small moat wall built atop of a '*que tai*' near the bank of a moat. Formally a temporary shelter for refugee livestock in time of war but later became fortified with loopholes, becoming a secondary defensive wall and an obstacle for the siege engines of the enemy. Also known as *'niu ma qiang'* (Cow and horse wall) or *'feng yuan'*.

Yarisama. (J. *yari*, spear). A loophole for thrusting pikes and spears. Typically, it is installed in narrow locations beside the main castle entrance and castle gates to skewer attacking enemy forces.

See *'sama'*, *'zama'*.

Yashikijiro. (J.). A small castle, more a country house with some defensive structures.

See *'jiro'*.

Yazama. (J.). An arrow slit common to Japanese castles, consisting of a narrow opening set in a defensive position in a tower or a plastered wall. The slit was splayed on the inside and narrow at the exterior, to allow the defenders a larger field of fire and to protect them from incoming fire. The slits had different names depending on the position they were used, seated (*'isama'*) or standing (*'tachisama'*).

See *'hako hazama'*, *'ishisama'*, *'sama'*, *'shinogi hazama'*, *'taihosama'*, *'tepposama'*, *'zama'*.

Yett. (Old E. *geat*, gate). An open-work iron grill suspended on hinges, placed on the inside of the wooden door of a main entrance to a tower; used as a secondary defence of the entrance. Examples can be found at Chipchase and Halton Towers, Northumberland, England, dating from the 17th century CE.

Yi cheng. (Ch. *wing wall*).

(1) As a walled city expanded in size, extra wall extensions were added extending from the *'lou cheng'*.

(2) The twin fortresses situated on the flanks and providing protection to a main fortress.

See *'cheng'*.

Yi Zi Cheng. (Ch. *one wall*). A section of wall extending from the main wall to a nearby river, or even sometimes into the water itself, which allowed access via a stairway or ramp to the river, for use inside the city for various purposes including domestic.

See *'cheng'*.

Yokobori. (J.). A horizontal trench across a gentle slope or mountain to increase the defensiveness of a mountain castle (*'yamajiro'*).

See *'bori'*, *'horikiri'*,*'tatebori'*.

Yokosama. (J.). A loophole for spears built into the narrow spaces beside main gates.

See *'sama'*, *'yarisama'*.

Yokoya. (J. *side arrow*).

(1) Structures for attacking the flank of an enemy with arrows, shot or ball, preventing enemy soldiers scaling a castle entrance or wall. Also known as '*sokusha*' or '*sokubou*'.

See '*sama*', '*zama*'.

(2) Castle structures built bent, indentured or curved including embankments, moats, compounds and towers, used for flanking attacks.

Yokoyahizumi. (J. *distorted rampart*). An angled projection from a castle wall to provide flanking fire.

See 'yokoya'.

Yokoyagakari. (J.). Projections of a castle's wall flanking fire can be provided, reducing the amount of '*dead ground*'.

See '*dead ground*', '*flanking*'.

Yokoya masugata. (J.). A defensive box shaped courtyard that projects out from the line of the walls of the castle. Provided with loop holes for archers and later gun loops for flanking enemy soldiers. Also known as a '*demasugata*'.

See '*masugata*', '*yokoya*'.

You no koguchi. (J.). An offensive entrance, set up for making sorties easier, with areas for mustering soldiers. Usually provided on flatland hill castles ('*hirayamajiro*') and flatland castle residences ('*kyojou kirajiro*').

See '*koguchi*', '*inno koguchi*'.

Yue cheng. (Ch. *moon wall*).

(1) An advanced work consisting of a semicircular fortified wall enclosing part of a moat or river, acting as a permanent bridge.

(2) A fortified wall enclosing the gatehouse of a '*weng cheng*', which were thinner and lower than the main city walls, allowing covering fire from the main walls.

See '*cheng*',

Yu qiang. (Ch.). The parapet on the rear section of a wall for safety.

See'*nu qiang*'.

Z.

Zahafah. (Ar.). Probably a fixed wooden tower for archers.

Zaqaziq. (Mameluke /Ar.). A spiked hurdle used in field fortifications.

See 'hersillon'.

Zangentor. (G. *pincer gate*). A German term for gate in a fortification which is set deeply between two inwardly angled exterior walls, forming a sunken road, which allows attack from the wall tops.

See 'gate'.

Zama. (J.). Loopholes in the walls of a Japanese castle of various shapes for firing guns and loosing arrows through at an enemy.

See *'hako hazama', 'ishisama', 'sama', 'shinogi hazama', 'taihosama', 'tepposama', 'yasama'.*

Zariba. (Abbasid).

(1) An Abbasid encampment surrounded by ditches and brushwood.

(2) An Abbasid woven wood field defence work for protecting archers.

Zasechnaya cherta. (Russian. '*Great Abatis line or border*'). A chain of fortification to protect the Grand Duchy of Moscow (13th to the 18th century CE) on the southern border, similar to the '*great wall of China*' or the '*limes*' of Rome. The border was made of '*abatis*' and ran for hundreds of kilometres (varied in length over time), with gates and watch towers.

See 'Ostrog'.

Zeughaus. (G. *armoury or arsenal*).

Zha ban. (Ch. *floodgate plank*). **See '*cha ban*'.**

Zha lou. (Ch. *flood gate tower*). A gate tower that contains the controls for the '*cha ban*' and sometimes the drawbridge, usually built on top of the gatehouses '*weng cheng*' or '*yue cheng*'.

Zhan peng. (Ch. *battle shield*). A temporary or semi-permanent defensive structure on the top of a wall, tower or bastion, as a hoarding, with a solid roof (made of clay covered in felt blankets or cowhides to absorb artillery.

Zharakty gate. (Otrar, Kazakhstan). A fortification commanding a key point connecting Kazakhstan with China, Europe, Near and Middle East, Siberia and Ural. A gate complex, with 1 metre (3 feet 3 inches) thick brick walls to a height of 4 metres (13 feet), surrounded by a moat, accessed by a bridge, with external and internal gates, guard rooms on either side of the complex, pointed merlons. Dating prior to the Mongol invasion. Provided with two types of embrasures;

rectangular with a small slit at the top, and rectangular with a curved apex.

Zhu fang cheng. (Ch. *garrison city*).

See *'man zhou cheng'*.

Zi cheng. (Ch. *child wall*). A defensive wall or an extension of the wall of an inner city.

See *'cheng'*.

Zig Zag. A trench excavated by besiegers and running towards the besieged position from the parallel trench in a zig zag pattern, to reduce the amount of exposure to defensive fire from the besieged.

See *'approaches', 'sap', 'trench'*.

Zinnen. (G.). Crenelations.

Zollburg. (G. *toll castle*). A German castle from the middle ages or early modern era, found mainly in the Holy Roman Empire, used to control the trade over alpine passes, border crossings and on the Rhine river, operating as a watch post or residence for the local lord or their delegate.

See *'castle'*.

Zone of contention. See *'no man's land'*.

Zone of fire. The area between the *'main line of defence'* and advanced skirmish or picket line, no wider than the range of the standard infantry weapons. Also known as the *'kill zone'*.

See *'no man's land'*, *'main line of defence*.

Zone of occupation. The area behind the *'main line of defence'* used for the logistics and support, including; trench systems, dug outs, supply caches, bunkers, magazines and covered ways.

See *'main line of defence'*.

Z-plan tower house. A tower house consisting of a rectangular block with towers protruding at the two diagonally opposite corners, a number of gun ports covered every part of the building's exterior wall, designed for the close quarters defence by musket and pistol.

See *'tower house'*.

Katiebell © 2024

Zugbrücke. (G.). Drawbridge.
Zwingburg. (G. *coercion castle*). A type of heavily fortified German castle built to dominate the surrounding territory, dating from mid to late Middle ages, built in areas where the locals were not in full support of the local lord.

See 'castle'.

Zwinger. (G. *keep*). An outwork consisting of an outer courtyard surrounded by curtain walls provided with flanking towers, forming part of a concentric fortification.
Zwischenwerk. (G. *intermediate work*). A small German fortified position that filled the gap between two main forts.
Zwischenwiderstandsnest (ZWN). (G. intermediate resistance nest). A German WWII stronghold built to fill the gaps in the '*Atlantic Wall*'

between the *'widerstandsnest'*, increasing the interlock and flanking fire.

See 'Atlantic wall', 'stützpunkt', 'stützpunktgruppe'. widerstandsnest'.

Zwingermauer. (*G. zwinger wall*). **See '*zwinger*'.**

Appendix 1.

What is a Castle?

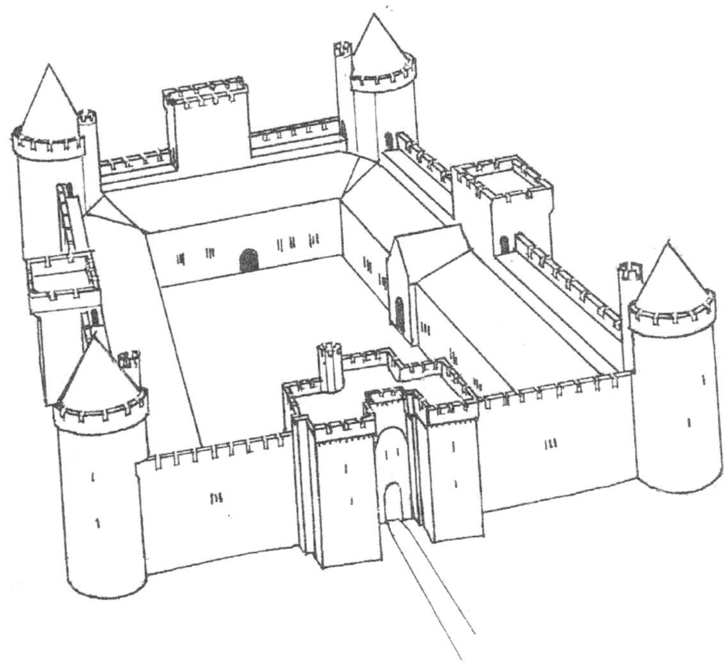

Figure 1. Quadrangular castle.

What is a Castle? Is it a masonry edifice constructed on a hilltop with tall towers punctuating the curtain walls topped by battlements? Can you imagine a column of knights in full plate armour mounted on war horses charging out of the main gate over the lowered drawbridge to do battle against an attacking enemy force, while beautifully coiffed ladies watch from a tower window, and serfs till the earth to grow harvest so that the lord of the castle can continue to spend money on courtly pursuits and new armour from Italy or Germany? This imagined spectacle represents the feudal system where the castle was a place of both residence and defence for a lord.

On the other hand was a castle simply a base for military and administrative operations, such as: dispensing justice and collecting taxes and tolls; subjugating conquered lands; suppressing rebellion and unrest; acting as a base for raids or invasions; protecting trade and territory from attack or just to attack another castle.

Whatever the case, the castle has become a visual symbol of the medieval period and everyone has their own view of the subject, be it coloured by their educational background and the bounds of their imagination.

As I have in fact studied the subject of military architecture to some extent I found that my definition of a castle is different to those less unacquainted on the subject. What are your ideas on the definition of a castle?

Some of the constructions that are given the name '*castle*' are in fact not castles at all. Oman I think puts it very well.
"*The word castle is linked to all manner of things, from prehistoric earthworks down to preposterous erections of the twentieth century, with mock battlements and sham drawbridges, built for misguided lovers of the romantic or for ostentatious profiteers.*"[2]

Since the Roman Juvenal called the British hill forts castella around 120 CE, such sites have been called castles, for example the multivallate hillfort near Dorsetshire in England is known to many as Maiden Castle.

And then there is the sad examples of Victorian and Modern architecture which grate on the eye nerve and make me cringe with disgust, especially when the owners make a profit from beguiling the unwary public and even receive grants from the government to do it (Bunratty and Kryal Castles (Australia) are two places that spring to mind).

The French use the word chateau for castle but it used to cover both. The Chateau-Gaillard, a fortress of the 12th century CE; and the manors dating from the 12th to the 18th century CE, which have no or little defensive capability.

[2] Oman, C., Sir. (1978). Castles, New York.

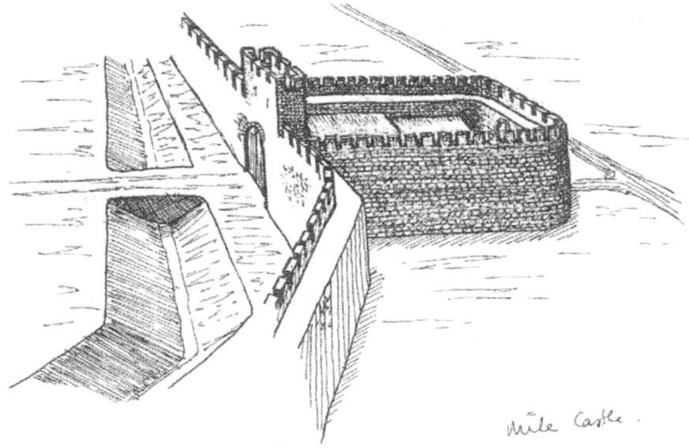

Figure 2. Roman Mile Castle.

The Roman Mile Castle (see Figure 2), which punctuate Hadrian's Wall are not in fact castles but towers defending the wall which forms a much larger fortification defending a frontier. If a Mile Castle is a castle then take the argument further, then every tower along the walls of Carcassonne in France is therefore a castle unto itself, which is an absurd proposition since the towers form part of a district circuit around a town and are too small to be considered a castle by themselves.

A fortified town, surrounded by curtain walls punctuated by flanking towers and other defences cannot be called a castle because of its large size. It would come under the category of '*fortress*'. Even though a castle may be attached or sited within the walls, a good example in Caernarfon in Wales, the castle is still a distinct unit unto itself, being able to cut off from the town by closing a gate or raising a drawbridge.

There have been numerous authors who have written books covering the topic of castles and other fortifications, and some of the more noted authors should have their definitions noted.

Sidney Toy, A History of Fortification from 3000 B.C. to 1700 A.D. wrote;
"*The Castle is considered in its military aspect, as a fortress, and its domestic arrangements only insofar as they are ancillary to its function as a fortification and are necessary in its residence.*" [3]

[3] Toy, S. (1955). A History of Fortifications from 3000 BC to AD 1700, London.

William Anderson, the author of "*Castles of Western Europe*" wrote; "*...The castle has a double nature; it is both a home and a fortress... [It is] this double nature which makes the castle so different from dwellings and fortifications of other periods.*" [4]

According to Sir Charles Oman in his book '*Castles*' stated that the following;

"*A castle is a military structure larger than a tower but smaller than a fortified town, it is residential but is also defensible in character. However complicated internally it must be a complete unit unto itself and not part of a town. A castle may exist inside a town but must be able to be cut off by the closing of a gate or the raising of a drawbridge.*" [5]

And finally the definition of a castle which appears in my own manuscript entitled "*A Dictionary of Military Architecture*",

"*A fortified building or set of buildings used to provide both active and passive defence. as well as a residence for the castle's lord and household.*" [6]

And in fact after researching and writing this piece I will have to change my entry for the definition of a castle.

All these definitions have some merit, but the point about the castle being the property of a single owner or feudal lord needs addressing. Castles were also owned or held by orders of knights. religious orders, republican states and commonwealths. And even the king of England did not own the Tower of London as a private person but as king. Therefore the question of ownership affecting the definition of a castle is removed, and the only conditions applying to the term castle are purpose size and character.

---

[4] Anderson, W. (1984). Castles of Europe, Hertfordshire.

[5] Oman, C., Sir. (1978). Castles, New York.

[6] Wyley, S.F. (1996). Dictionary of Military Architecture.

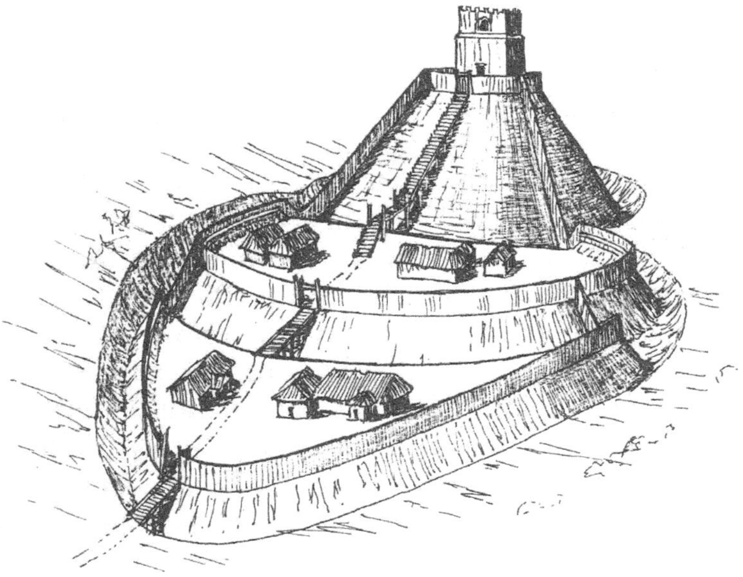

Figure 3. A Motte and Bailey castle.

Over the Medieval period a number of different types of castle designs were used, such as the motte and bailey castles (see Figure 3), the keep and bailey, using a range of keep designs (e.g. shell (see Figure 4), rectangular, square, quatrefoil, round and polygonal keeps), quadrangular castles (see Figure 1), keep-less castles, and the concentric castles. There were certainly other types and variations on themes coming under the definition provided in this text.

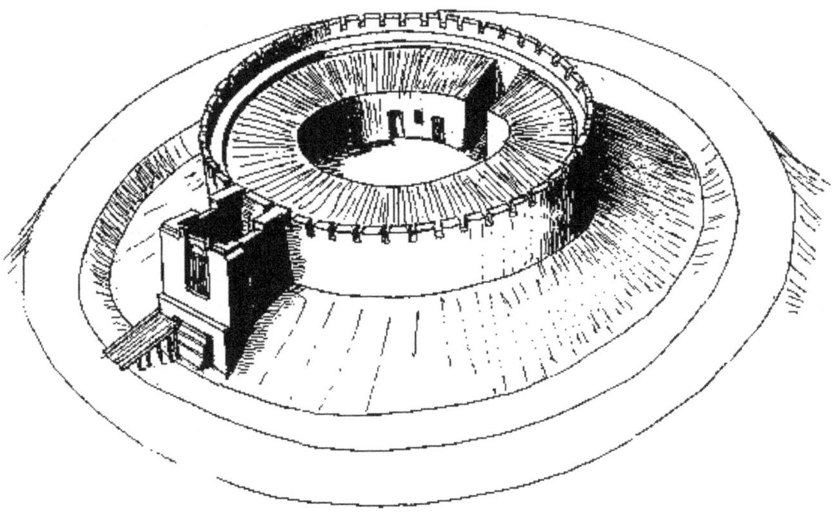

Figure 4. A Shell keep.

The history of the word castle is worth noting because of the confusion created by both Medieval chroniclers and Modern historians in their mistranslations of such words as castrum, castellum and other similar words. Castrum vaguely means any enclosed place such as a fort or stronghold of any type. The plural of Castrum is Castra, which was used to denote a legionary fort, and castellum is the diminutive of castrum. So when Wycliffe translated the Latin bible into English he thought castella was a little town. When the Vulgate translation was made in the early fifth century CE castellum and castrum were still being used in a vague sense, both being applied to centres of population either big or small, fortified or not. When St Jerome translated the New Testament into Latin he used Castellum to mean a village of any type so when the author of Cursos Mundi about 1300 CE, stated that Bethany was the castle of Lazarus, Martha and Mary, in which they entertained Jesus, he was thinking of a fortified residence.

In Carolingian France and Anglo-Saxon England in the eighth and ninth century's CE castellum was used to mean a habitation of any size, for example Egbert of Kent and Offa of Mercia both called Rochester a castellum, and the sizeable towns of Macon and Vitry were called castella by a Frankish chronicler. It was not until the tenth century that the word actually began to be used in its actual sense of a fortified residential building. Since no new word was coined to cover this new phenomenon this confusion arose, and thus both medieval and modern historians have misunderstood and passed on that misunderstanding to others.

In the transition period between the senses in the tenth and eleventh centuries CE it is impossible to discover whether a writer was writing about a town or a fortified stronghold, except from the context in which it was written. The phenomenon of the castle spread throughout Europe and beyond, and many variations of the word were derived, such as Chateau, Kastel, Castello, Castillo. Kastros, Kastris and Castelle.

Appendix 2.

Shiro - A Japanese Castle.

Introduction.

A castle is a fortified building or set of fortified buildings used to provide both active and passive defence, as well as a residence for the castle's occupants, usually the castle's lord and household. A '*shiro*' is the Japanese equivalent of a European castle, a '*shiro*' may appear to European's eyes as something far less robust to what they would be used to in their homelands but a '*shiro*' fulfilled the same functions as their European counterparts. The aim of this essay is to examine a '*shiro's*' design features and their functions.

Himeji Castle, Hyōgo Prefecture.

The building we know today as a '*shiro*' (or '*jō*') was developed in the short period between the last part of the 15th century and the first part of the 17th century CE. The major elements of the design and technology were borrowed from both religious and residential architecture of the period.

Oda Nobunaga (1534-82 CE) had *Azuchi - Jõ* built between 1576 and 1579 CE, marking his rise in the struggle for control of Japan in the 16th century. Many other lords followed his lead, each lord needed a strong defensive base to support his aspirations. But once the forces under Tokugawa Ieyasu (1542-1616 CE) finally defeated those of Toyotomi in 1615 CE the age of the '*shiro*' was at an end. In the sixth month of 1615 CE a shogunal edict was issued ordering the destruction of all regional fortifications except for the actual residences of the local lord. A document from the period stated that four hundred castles and lesser fortifications were dismantled in only a few days.

Strategic Placement.

'*Shiro's*' were built on strategic sites to take advantage of the position itself (eg. height advantage, difficult access for an aggressor, water and supply availability, ease of communication and control of transport routes).
The three main situations '*shiro's*' were built were;

1) on the top of a hill ('*yamajiro*');
2) on a hill surrounded by a flat plain ('*hirayamajiro*'), and;
3) on a flat plain ('*hirajiro*').

A '*yamajiro*' (J. mountain castle) was usually built on a mountain top or some other elevated position, constructed using walls of rocks and earth to reinforce the natural defences of the site, a moat was optional. This form of castle was the favoured form prior to the latter half of the 16th century CE, where the lessening of civil strife began to show an effect.

A '*hirayamajiro*' (J. flatland hill castle) was constructed on a hill on a plain, thus taking the positional advantage of the surrounding area.

A '*hirajiro*' (J. flatland castle) was developed after the necessity of the '*yamajiro*' fortification declined with the cessation of civil war in the 17th century CE. These castles were built on the plains and served as administration centres of the surrounding area.

Fortified compounds.

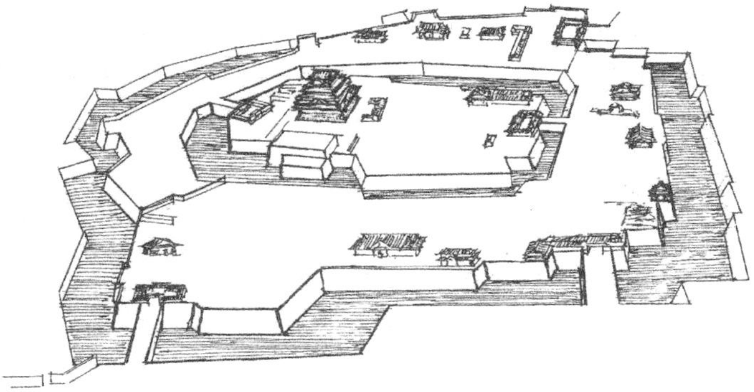

Figure1. *'Honmaru'* and *'ninomaru'* of *Osaka – jõ.*

Most *'shiros'* consisted of either one or more joined compounds or '*kuruwa*', these were the:

1) *'honmaru'*,
2) *'ninomaru'*, and
3) *'sannomaru'*.

The *'honmaru'* (J. main circle) (see fFigure 1) was the main or inner fortified compound of a *'shiro'*. The *'ninomaru'* (J. second circle) was the second defensive compound of a *'shiro'*, which either surrounded or was adjacent to the *'honmaru'*. The *'sannomaru'* was the tertiary defensive courtyard of a *'shiro'*, depending on the design, the *'sannomaru'* either surrounded the *'ninomaru'* or was sited adjacent to it.

A *'dôshinén'* was one of the three major types of *'shiro'* which date from around 1600 CE. A *'dôshinén'* was constructed using a plan of three fortified compounds arranged in concentric pattern; the *'honmaru'* in the centre was surrounded by the *'ninomaru'* which in turn was surrounded by the *'sannomaru'*, variations on this occurred depending on the site.

The *'renketsu kei'* was another one of the three major types of *'shiro'* which dates from around 1600 CE. Arranging the *'honmaru'*, the *'ninomaru'* and the *'sannomaru'* in series along an axis, and connecting them by gateways formed a *'renketsukei'*.

The '*teika kukei*' is the final one of the three major types of '*shiro*' which date from around 1600 CE. The '*teikakukei*' consisted of two defensive compounds; the inner or '*honmaru*' was set to one side of the encircling secondary compound or '*ninomaru*'. This positioning of the '*homnaru*' was to remove it from the most likely line of assault, thus gaining extra protection.

Basic Design.

The basic design of a '*shiro*' consists of four main elements.
1) Moats or '*hori*';
2) Walls or '*ishigaki*';
3) Gates or '*mon*';
4) and Towers or '*tenshu*'.

***Ishigaki* and *Dobei* (walls).**

Figure 2. '*Yagura' is atop an 'ishigaki'*.

An '*ishigaki*' or dry stone wall (see Figure 2) was used in the construction of the main walls of a *shiro*. The '*ishigaki*' was introduced into Japanese military architecture because of the increase in the frequency of conflict and earthworks were no longer a form of effective defence. The technique of construction involved embedding stones into an earthen embankment and locking them in place with smaller stones. This allowed the whole wall to shake and thus dissipate the effect of earthquakes, which are a normal occurrence for the country. The *'ishigaki'* were used as a foundation for such structures as; *'tenshus'*, corner towers, parapets and connecting towers.

A *'dobei'* was a wall made by daubing clay over a reed or bamboo framework and topped by a small tiled roof. This type of wall was used in two different parts of a *'shiro'*.

They were built along the top of the *'ishigaki'* (or the main walls) linking the corner and connecting towers, providing protection to the defenders within;
and the *'dobei'* were also raised on both sides of the castle's major ways, effectively screening the rest of the compounds from enemy eyes, and they formed a maze which hindered the enemy greatly if they gained entry.

***Mon* (gates).**

Figure 3. Masugata mon in the centre and kôrai mon on the left.

A *'watari yagura' mon* was the largest kind of gateway used in a *'shiro'*, which consisted of a two storey timber frame structure, where the lower storey incorporated the entrance, and the upper storey formed the connecting tower (from where the gateway derives its name). The

style of the connecting tower is typical of a *'yagura'* (see Figures 2 and 3) but the lower storey's timber is left unplastered, and the actual gate and surrounding timbers were often covered with strips or sheets of iron which improved their protection against battering and firing.
The *'kôrai mon'* (see Figure 3) was the smaller initial gate of a *masugata mon* or barbican gatehouse complex of a *'shiro'*, consisting of a three roofed structure (a central roof was flanked by another two roofs). This gate restricted access to the courtyard beyond, and if the enemy did gain entry to the courtyard they would be confined within that space and would be exposed to the fire of the defenders.

***Tenshus* (towers).**

Figure 4. Tenshu.

The main tower of a *'shiro'*, constructed mainly of timber on a base consisting of *'ishigaki'* or earth revetted with dry stone walling, serving a number of purposes such as; ceremonial, residential and military. The early *'tenshu'* were simple residential buildings with belvederes perched on their rooftops, later they developed into the larger multi-levelled towers.

As *'tenshu'* (see Figure 4) developed various types evolved, they were:

- the *'boroshiki tenshu'*, which was the earliest form of *'tenshu'*, consisting of a multi-storied timber, framed tower with a belvedere or observation tower set on top of the roof ridge;
- the *'renketsu shiki' tenshu*, which was the most ornate and complex type of *tenshu*, the *tenshu* was connected to smaller towers by parapets at the corners of the rectangular stone base; the *'renritsushiki tenshu'* which was attached to a secondary tower by a connecting parapet;
- and finally the *'teiritsushiki tenshu'*, which consisted of a principal tower, attached directly to a secondary tower, was also known as a *'fukugôshiki tenshu'*.

Other design features.

Various other features of a *'shiro'* went towards making it very distinctive, if not more defensive in nature.

Dôzo-zukuri.

Since most of the buildings of a *'shiro'* were mainly constructed of wood there was the necessity of reducing the risk of fire. This was accomplished by the technique of covering the walls and timbers with thick layers of plaster, which is known as *'dôzo-zukuri'*. The plastering was used as a protection against domestic fires as well as those caused by the incendiary missiles of an enemy.

Ishi-otoshi mado.

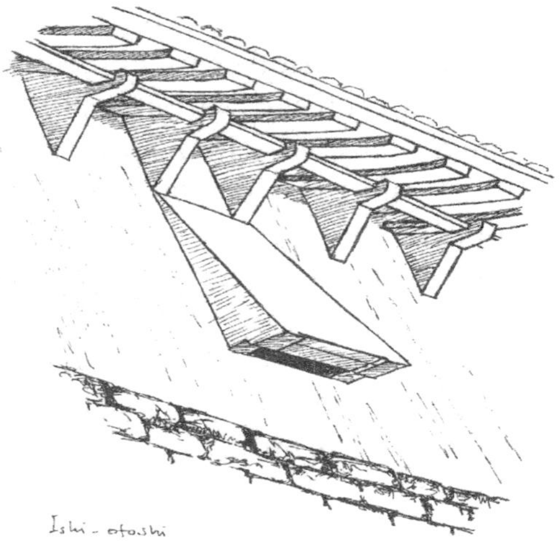

Figure 5. Ishi-otoshi mado, stone dropping window.

An *'ishi-otoshi mado'* (J. stone dropping window) (see Figure 5) was structure which was used for the vertical defence of the walls of a *'shiro'*, consisting of a section of the wooden or plastered wall of a; *'tenshu'*, tower, connecting tower or other structures, which was built out over the stone walls at strategic locations. The structures enabled the defenders to shoot arrows as well as throw stones at attackers attempting to scale the walls. The aperture could be closed for protection if necessary by a panel of thick wood that was strengthened by iron bars.

Teppôsama.

A *'teppôsama'* or musket port was set into the plastered wall or tower of a *'shiro'*. In section a port was funnel shaped, the larger part on the inside gave a greater field of fire to the defenders, while the narrower aperture on the exterior provided protection to the defenders from incoming fire. The *'teppôsama'* came in a number of shapes; rectangular, triangular and circular. The ports of the plastered walls on the *'ishigaki'* were left open while those of a wall tower were often provided with shutters, since they were put to other uses other than defence (e.g. storehouses or residences).

Conclusions.

'*Shiro*'s' were built to provide a logistics base from which a lord could use to facilitate his campaign. Without a secure site for storage of supplies and defence in case of an attack the lord could never fulfil his grand designs.

The architectural design of a '*shiro*' utilised the materials, principles and design features already used for domestic and religious buildings. With the addition of a few defensive design features the shiro became the architectural and historical monuments of the period.

Only a few extant castles survive to this day out of the hundreds that were built in a period of strife. This is due to two major factors. 1) the Shogunal edict of 1615 CE; and 2) the nature of the materials used in their construction.

Appendix 3.

Siege Warfare - The Art of Offence and Defence.

Introduction.

This article was first presented as a lecture at the Fifth Australian Medieval Conference, (*Four Winds, Maldon, Victoria*) in 1989 CE.
The development of fortification has always been partly dictated by the type of siege tactics and equipment used against them. The basic techniques of siege craft established in the ancient world continued to dominate siege warfare until the introduction of the cannon; some techniques were still employed in this day and age.

Basically, an attacker confronted by a wall has five options, if you include retreat. The are as follows;
1) Blockade – to cut off supply of provisions and reinforcements to the fortification.
2) Escalade – to attack the wall using scaling ladders and or siege towers.
3) Breach – to cause a breach in the wall by attacking the masonry of the wall using siege equipment (bore, ram, mouse).
4) Mining – to cause a breach in the wall by undermining it, or combination of escalade, breaching and mining.
5) Retreat – to retreat was also an option when attacked or threatened by a relieving force.
(6) Trickery- to gain access by means of ruse or treason (i.e. Saracen capture of Krak des Chevaliers, Syria).

Blockade.

An attacker may take the first option and wait for the garrison to succumb to disease, starvation, thirst or boredom. Such tactics could not be relied upon because if the garrison's larders were well stocked and the supply of fresh water was adequate, the besieging force had a while to wait, during which the besiegers had to survive on what they brought with them or off the surrounding countryside (not a good option if the besieged had implemented a scorched earth policy).

Escalade.

Escalade was the most direct and also the most dangerous of the options. The action consisted of the attacking forces moving forward from their lines to the base of the wall where they set the ladders against the wall, and then scaled them to engage the defenders in hand to hand combat. The attacker had to reach the wall top while being harassed by defenders trying to push the ladders away, throwing missiles and loosing arrows and bolts at them.

Eventually, if the attacking force was lucky enough they made ground on the wall top by being continually reinforced from below, they could force their way into the castle, opening the gate for the rest of their companions left waiting outside. This was not always a successful option due to the fact that the defending garrison had their own ideas on the matter.

Even if the attackers did manage to gain control of the battlement, any further advancement was usually hindered by the fact that the defenders could retreat to the towers punctuating the curtain wall thus isolating the wall walk, and continue the defence from the adjacent towers. The attackers would find it a difficult position to hold, being harassed from the inner works as well as the flanking towers, and that any siege equipment used to breach the towers would probably have to be brought up the ladders, not to mention the exposed nature of the position.

A siege tower was a much more efficient way of attacking a wall top position, allowing a large fighting force to gain the wall top in a short period of time in comparative security compared to an escalade.

A siege tower usually consisted of a wooden tower on wheels or rollers, equipped with ladders, stages and drawbridges at the top. The exterior could be clad in plank or hides to ward against missiles and incendiary devices. The tower was built to the height of the wall, filled with fighters, and then pushed into place next to the wall, then the drawbridge on the top deck was lowered and the fighters rushed

forwards to engage the defenders, while reinforcements moved upwards from the lower stages adding to the weight of the initial assault. Siege towers often were defended by archers shooting through arrow slits or crenels, and also equipped with sally ports and even a ram in some cases.
The Siege tower could easily be countered by surrounding the fortification with a ditch or a moat. The attacker overcame the problem by filling in the ditch with bundles of brushwood, rocks and soil, producing a causeway towards the wall. Defenders could also dig pits covered with brushwood and earth outside the walls to counter an attack by siege tower, which would either be stopped or would topple over when they ran into them. Defenders would also be retaliating during the advance of the siege tower itself and its occupants, and the forces moving the siege tower into place.

Breaching the walls.

The wall masonry could be attacked directly, breaking a breach through which the assailing forces could enter the fortified perimeter. One way of causing a breach in the stonework was to use a heavy iron-tipped ram to pound the wall until it gave way.

The operators of the ram were under attack most of the time while they were at work from above, the defenders hurled all manner of objects and materials in an attempt to impede the work. So it was always wise to employ a penthouse or 'cat', a wooden shelter protected by rawhides or iron plants. The ram itself was slung on chains beneath the penthouse's roof, once the penthouse was set in place against the wall, the ram could be set swinging against the masonry. Another device which could also be used under the penthouse was called a 'mouse' or a 'bore', which was a sharp tipped lever which was used to pick out the mortar between the stones and eventually the stones themselves, until a breach was made.

Mining.

Mining was one of the best ways of breaching a wall, though it was a slow process. A tunnel was dug from the besieger's lines to under the wall to be breached, once under the foundation, the tunnel was extended laterally along the wall front. The chamber was supported by timbers and once it was of a sufficient size, it was filled with combustible material and then set a light. Once the wooden props had burned away the section of wall above was thus unsupported and gave way falling into the cavity created by the mining, producing the required breach.

Mining could be obstructed by making the base of the walls wider (e.g. Spurs, talus, etc.) making the wall more stable. The only active countermeasure to mining was counter mining. A mine was excavated from inside the castle to intercept the incoming mine. The combat which occurred when the two mines met in those dark, cramped conditions was one of the most horrific aspects of early siege warfare.

Siege Engines.

In addition to the direct methods of assault, there were the engines which acted as artillery, which were capable of casting missiles over the walls of fortifications and causing significant destruction. There were two basic types; the mangonel, which was a catapult; and the ballista, a massive crossbow.

It was not until the Middle Ages that a new siege engine was developed, the trebuchet. This consisted of a long pivoted arm. At the front, short end, a basket of stones acted as a driving force, at the longer end was leather sling. To operate the device, the long end of the arm was pulled down against the weight of the stones and missiles were placed in the sling. This was usually a stone but artillery men have always had a morbid sense of humour, and dead horses, corpses and even live prisoners found their way into the sling at various times. The arm was released, the counterweight causing the front end to go down and the long arm to fly through the air and discharge the contents of the sling in a high arc which carried the missile over the defences.

Although engines were often used to batter walls by throwing massive stones against them, their prime purpose was to bombard the interior of a defended work or town so as to wreck buildings, kill or injure and demoralize the occupants. Incendiary missiles such as red hot stones or jars of '*Greek fire*' were also used. Siege engines were also utilised by the defenders against the attackers, the advantage of height being in the defenders favour. Introduced by the Greeks and perfected by the

Romans, siege engines continued to be used until well into the gunpowder era, since they had virtues of being cheap and easy to make and were much more destructive than early cannon.

Conclusion.

To conclude, siege warfare was a time consuming affair which pitted the strengths of military architecture against the destructive effects of siege craft. Overall, the advances in siege craft were countered by subsequent advances in military architecture, and the reverse.

S.Wyley 18/3/89.

Appendix 4.

Design, build and maintain your own European castle.

By Stephen 'Sven' Wyley.
18/10/2024.

Introduction.

So you want to design, build and run our own castle, well there are many considerations such as; location (in regards to access, is it going to be coastal, inland, riverside, close to transport, close to materials, and what about site viability), purpose (defensive or offensive), design (overall type versus selection of shape of crenelations), legal, political and social issues, the costs of materials, labour, tools, let alone an architect, then once is built the ongoing maintenance and repair. Access to resources for building the castle, such as stone, timber, mortar, water, iron, could increase the cost if they had to be transported long distances.

As to a definition of what a castle let's go with: '*a castle [is] simply a base for military and administrative operations, such as: dispensing justice and collecting taxes and tolls; subjugating conquered lands; suppressing rebellion and unrest; acting as a base for raids or invasions; protecting trade and territory from attack or just to attack another castle.*'[7]

A great example of experimental archaeology is Guédelon Castle, a castle based on 13th century design which is currently under construction near Treigny, France.

Time period.

For the purposes of this article let's keep the period to the medieval period (500 to 1500 CE).

---

[7] Wyley, S. (2024)

"Over the Medieval period a number of different types of castle designs were used, such as the motte and bailey castles, the keep and bailey, using a range of keep designs (e.g. shell, rectangular, square, quatrefoil, round and polygonal keeps), quadrangular castles, keep-less castles, and the concentric castles. There were certainly other types and variations on themes coming under the definition provided in this text."[8]

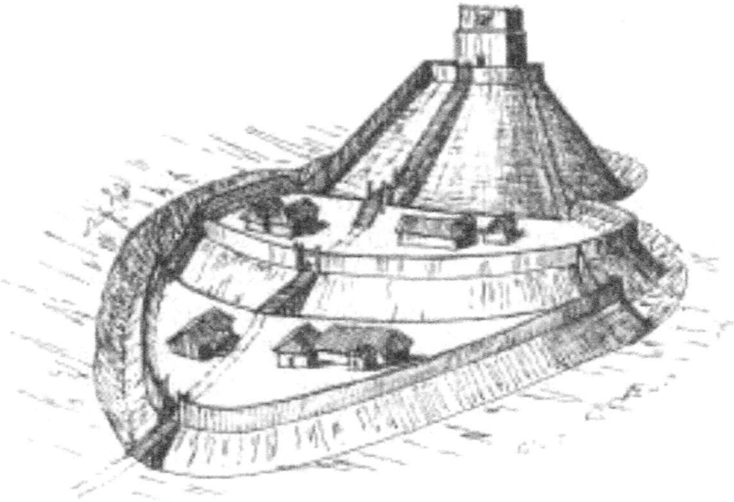

Drawing 1. A motte and bailey castle, with an outer and an inner bailey.

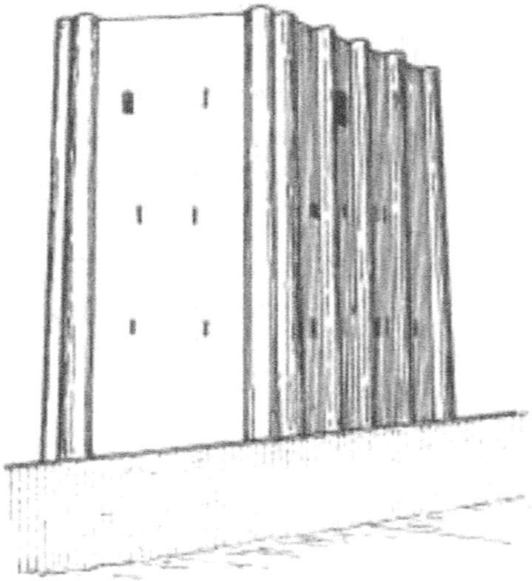

Drawing 2. A tower keep.

[8] Wyley, S. (2024)

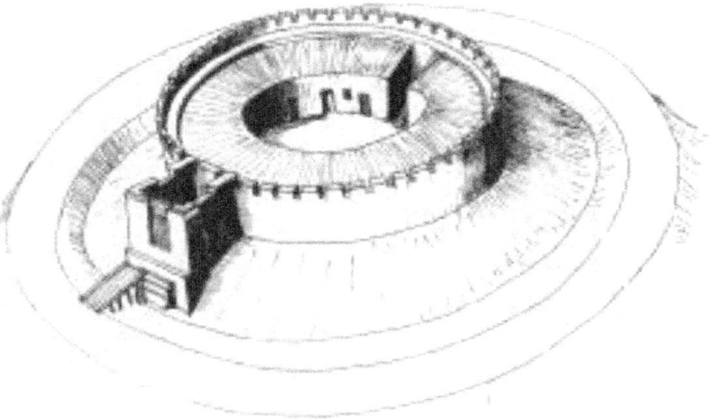

Drawing 3. A shell keep.

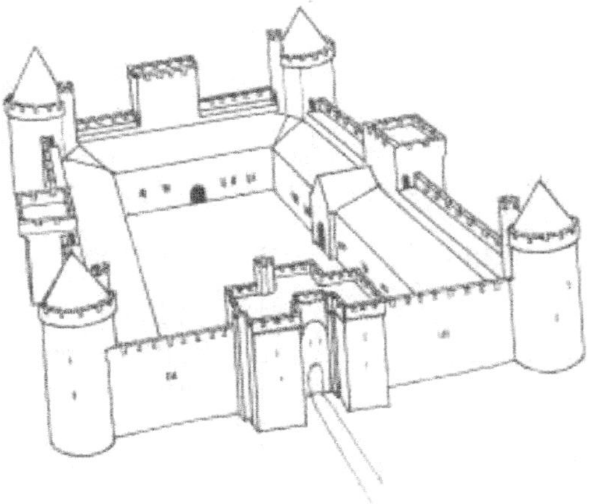

Drawing 4. A quadrangular castle.

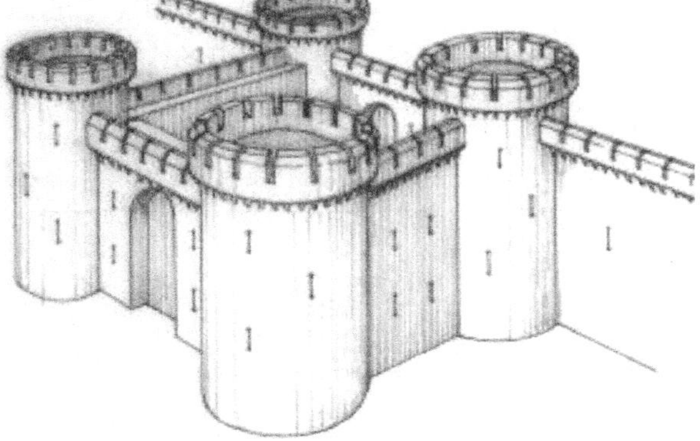

Drawing 5. A great gatehouse.

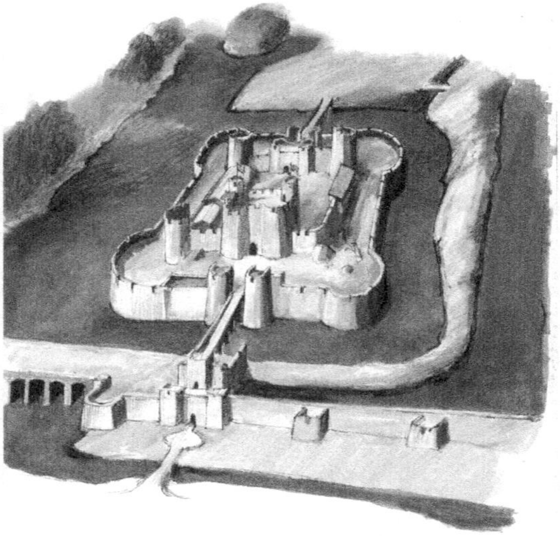

Drawing 6. A concentric castle.

Timeline of development.

Type of castle	*Date (CE)*	*Example*
Motte and Bailey	9th century	Brinklow, Warwick , England, UK
Keep and bailey	992	Langeais, France.
Shell keep	First half of the 12th century	Restormel castle, Cornwall, UK.
Keep with fore building	12th Century	Rochester Castle, UK.
Concentric	1267-1277	Caerphilly Castle, Wales, UK
Keepless (Great gatehouses	1285-1290	Harlech castle, Wales, UK

Location.

A castle could be built on flat land or various high ground, whether it is on a mountain top or mound such as a motte (as in a motte and bailey castle).

Access to water for a castle was important for two reasons:

1) Access to a harbour on a waterway or the sea coast allowed access for resupply and trade.
2) Clean drinking water was important for life but was also important for ablutions, the making of mortar, quenching metals etc.

Climate and weather conditions need to be taken into consideration, for example in arid zones water is valuable so infrastructure is required for the collection and storage of rain water.

There are a number of geotechnical issues that affect a castle site, including the height of the water table, soil or rock type, and the effects of erosion. So cliff tops overlooking a sea can be defensive, picturesque but erosion and land slippage will put a hole in your dream.

Castles could be built on previously fortified sites (Roman or Iron Age hillforts), waste not want not, the ditches, walls let alone the stones themselves can be re-used or built upon.

Access to building materials (stone, brick, tile, slates, timber, iron, water and lime) is important for building a castle affecting availability and cost of transporting such materials.

Digging moats and raising earthen ramparts does not take a lot of skill but to build a castle you need skilled stone masons and carpenters which will have to be sourced from, in some cases, overseas, and thus will add to your costs.

Size.

The size of your castle is affected by the size of the site available, the design and materials available, the amount and skill of labour, and the amount of money available. The bigger the castle the higher the costs, the bigger the perimeter the more soldiers required to defend it (more food, fuel, weapons etc. require.

The size and complexity of a castle also affected how long it took to build, lets looks at Conway castle (Wales, UK), it had 1280 metres

(1400 yards) of curtain wall some 9 metres (30 feet) high, 22 towers, and three gates, and took 4 years to build, built between 1283 and 1287 CE, and of course build had to stop during December and March because of the poor weather.[9]

Development over time.

Ditches and hedges were used in the Iron Age to denote boundaries and could be used in defensive works (see Offa's dike), the same could be used for the defence of smaller areas such as a homestead. Later including a wood palisade and gate tower, then tower for refuge...

Parts of a castle.

Depending on the time period castles would have various parts required for defence and offence; including moats, ditches, walls, towers, fortified gates, drawbridges, murder holes, sally ports, battlements (crenels and merlons), machicolations, arrow slits and gun ports.

Along with all the military stuff there has to be room for housing, the armoury, the stables, storage areas like the cellar, the cook house to provide food for the occupants, the guard rooms (in the gatehouse), the barracks for the rest of soldiers, various interconnecting galleries, stairways, passageways to get from one point to another, and then there was the chapel, the well, the hall, the fireplaces, and latrines (garderobe).

Materials.

Dirt is *'dirt cheap'* and should be available for most sites to raise ramparts and mottes. Whereas suitable clays to make tiles are harder to find. Tiles can be made on site for roofing and flooring can be found and fired on site if you have someone onsite who knows how to build and run a kiln. Similar will be the glass for windows, a most expensive item. Whereas timber grows on trees, and should grow in most areas, oak is recommended but not available in all areas. Timber is required for palisades, gates, doors, portcullis, brattice and hoarding, roof and

---

[9] Hindley (1968).

flooring, furniture. Suitable stone may be available at site but it requires specifically training people to quarry, cut and fit stone to make a castle, adding to the costs. The mortar used to hold the tiles and stones in place requires lime, sand and water. The sand and water are the easy part, the lime requires processing from limestone or shells in a lime kiln, heated to 1100° Celsius (2000° Fahrenheit). Roofing can be made from thatch (bundles of reed), shingles and shakes are made by splitting timber into thin boards, slate tiles for roofing are split from quarried slate, and roofing tiles are made from clay and fired in a kiln at 1100° Celsius (2000° Fahrenheit). Wooden pegs used to hold frames together are made from timber which requires splitting down to size with a froe and shaping to size and shape with a drawknife. The iron for nails needs to be mined, then smelted 1250° Celsius (2282° Fahrenheit) to remove the impurities, then forged to length and shape and the head shaped.

Labour.

All the preparations of the site and materials takes labour, time and money, the same with the erection of the castle including:

- diggers (earth moving);
- carters (moving the materials);
- timber cutters (to cut the timber, to split the shingles and shakes, for the carpenters);
- carpenters (to build the woodwork for the drawbridge, the gates, the portcullis, the flooring and walls and doors, the brattices and hoarding);
- rope makers,
- miners (to mine the ore for the smelters);
- iron smelters (to smelt the ore for the blacksmiths);
- blacksmiths (tool makers, nail making, fittings (hinges, grills, locks, portcullis, door studs, yetts, ;
- quarrymen (quarrying the stone for the masons);
- masons (cutting and fitting the stone work together);
- master mason (overseeing, design, engineering and architectural services);
- clay diggers (to collect the clay);
- tile makers (to shape and fire the clay into floor and roof tiles);
- lime burners (to turn the limestone into lime);

- charcoal burners (to turn wood into charcoal for the fires of the blacksmiths). And the list goes on...

Legal issues (licence to crenellate).

In England a person required a warrant of royal permission to build a castle or to fortify an existing structure such as a manor house. If you did not have a licence to crenellate it is known as an '*Adulterine castle*'. The unlicensed castles (probably of the 'motte and bailey' design) were constructed without royal permission by the Barons supporting Matilda in England during the reign of King Stephen (1136-1152 CE), and demolished by King Henry II after the treaty of 1153 CE. An example was King Stephen's brother Henry, Bishop of Winchester who had to destroy the three castles he had built near the cathedral city.

Political issues.

Castles were centres for control over the countryside and people for the purposes of taxation, administration, barracks, movement and rebellion control.

After 1066 CE the Normans used '*motte and bailey*' castles to subdue the Saxons. King William built the first stone tower replacing timber and earth construction in 1070 CE(the Tower of London). The consolidation continued and between 1070 and 1087 CE another eighty five stone castles were built across England.

The castles built under the direction of Edward the 1st in Wales were the pinnacle of design and cost a fortune but delivered control of the Welsh.

Running, Maintaining and repair.

Running a castle was not cheap. During peace a castle would only need a castellan to manage the castle and a small garrison on troops for guard duty. During a siege the garrison would swell in size, to win a siege the attackers usually needed about 3 to 1 advantage to win in numbers of troops.

During an attack on Harlech Castle (Wales, UK) in 1294 CE by Prince Madog ap Llewellyn (Welsh) the garrison of just thirty seven soldiers repulsed the attack.[10] Whereas the full strength garrison of Krak de Chevaliers (Syria) when held by the Templar knights was about 2000 soldiers.[11] [12]

All the staff needs to be fed and clothed. Timber will be needed for repairs (timber is attacked by insects, flammable and susceptible to rot). Timber will also be needed for tool handles, furniture, arrows, bows, crossbow stocks, looms, buckets, fire wood. Other materials will be required for the manufacture of goods, such as: leather, feathers, horn, bone, wax, dyes, and fabric.

All buildings decline with time, timber gets borer or is exposed to moisture and so rots, fires can occur burning all the combustible material and damaging stone. The wooden towers of the Normans on top of the mottes were susceptible to fire and rot but only a few were replaced with stone tower on top. And moats fill up with debris and silt and so need to be cleared on a regular basis.

Cost.

Castle building, garrisoning, maintenance and repair was usually never cheap, someone had to pay, normally the tax or toll payer. There are a number of references to the costs involved in the historical sources.

William the 1st and his successors took advantage of pre-conquest laws providing for the repair of burhs to force the locals to provide labour to build castles (motte and baileys mainly) which was a cheap way to get a castle built while oppressing the people.[13] Henry II's update of Dover castle cost a total of £7,000 ($13, 646 AUS), including the keep, curtain wall and twin towered gateway.[14] Edward the 1st conquest of Wales saw ten castles built from Aberystwyth to Rhuddlan, including the best examples of concentric castles and great gatehouse, which cost the

[10] Hughes (1974).
[11] Hindley (1968).
[12] For more on garrison size see Bradbury (1992).
[13] Kightly (1979).
[14] Kightly (1979). $AUS as of 17/10/2024.

equivalent in modern terms to £10,000,000 ($194,950,000 AUS).[15] Harlech castle, one of Edward the 1st castles in Wales took seven years to build with a wonderful example of a great gatehouse, an equivalent of a keep, at one time in the construction employing '*227 masons, 115 quarries, 30 smiths, 22 carpenters and 546 labours*', and 26 administration staff[16] , costing in modern terms £2,000,000 ($3,899,000AUS).[17] 300 '*fossatores*' (i.e. navies or construction workers) were sent to work on the castles of Flint and Rhuddlan in Wales from Hoyland in Yorkshire in 1277 CE.[18] Bolton castle (a courtyard castle), commenced building in 1377 CE cost about the same as Harlech castle.

Changes in warfare.

Once you have built your castle you will need to update design to counter the changes of warfare including how castles are attacked and defended, including:

the change in artillery from trebuchet and ballista to cannon and mortar which saw the walls getting lower and thicker, the shape of battlements changing from upright to curved to deflect incoming cannon wall, outworks will need to be added to increase the distance between the castle, under mining changed from firing the supports to using gunpowder to blow up a wall to make a breach, hence the development of talus and spurs (which improved the stability and strength).

Conclusions.

In conclusion, a castle is a geostationary defendable point of control for a lord or king, taking advantage of defensibility of the location (be it natural and person made). The site must be defendable for given amount of time before the food and water run out (or no one is coming to break the siege), given the location, design, the amount water and stores, number of defenders available, the type of weapons and number of attackers brought to be bear (ladders, siege towers, artillery

[15] Kightly (1979). $AUS as of 17/10/2024.
[16] Hindley (1968).
[17] Kightly (1979). $AUS as of 17/10/2024.
[18] Hindley (1968).

(trebuchets or cannon)). The design of a castle should reflect the most up to date type and reflect the current best practice. The highest grade of materials should be used, the highest skilled workers for the best price is the ideal. Good luck in your endeavours.

Appendix 5.

Linear fortifications.

Introduction.

Long fortifications which could take various forms varying from a simple ditch and rampart to heavily fortified frontiers.

Looking north from the Yedicule at the land walls of Constantinople (1992).

Linear fortifications were static land based defences that needed to be defended, maintained and where necessary upgraded, they were expensive to build and maintain, their longevity can be overcome by changes in circumstances (the fall of the Roman empire, *'Hadrian's Wall', 410 CE*), changes in politics or borders (*'Danwerk'*, 650 CE to the 19th century CE), changes in types of warfare (German *'blitzkrieg'* of Poland, 1939 CE) or by passed (Germans attacking the French via Belgium bypassing the expensive and well fortified *'Maginot line'* 1940

CE), they could be overcome by an overwhelming force (the Allies' Normandy landings, '*Atlantic Wall*', 1945 CE) or by changes in fire power (the cannons of the Turks, '*Theodosian Land walls*', 1453 CE).

Definitions.

For the purposes of this article here are a number of definitions of the linear fortification used in this research. You may not agree with the definitions used but it is a good place to start.

Bastioned enceinte. The defences of a town/city formed by a bastion enceinte (Floriana lines, Malta).
Border. Land border between two or more countries (Alpine wall, Italy).
City walls. The fortifications defending a city (Constantine wall, Constantinople).
Coastal. Along a coast (Atlantic wall).
Fort. A single fortified site (Great wall of Sindh, Talpur dynasty, India).
Frontier. The contested zone between two nations, which can be a stop line (Ukraine vs Russia).
Harbor defences. Fortifications defending a harbour (Corradino Lines, Malta).
Isthmus. Cutting of access to a neck of land (Hexamilion wall, Isthmus of Corinth, Greece),
Peninsular. Cutting of access to a peninsular jutting out into a body of water (Lines of Torres Vedras, Portugal).
Stop lines. Fortified lines to stop or hinder the advance of an enemy (Gustav line, Italy).

Of all the types of linear fortifications looked out in this research the borders were the most prolific, followed by capital defences, then stop lines and peninsula.

The French and Germans owe their defeats in Russia to the Russians willingness to cede ground, even a capital, with their enemies long supply lines coupled with a scorched earth policy, it was a lost cause before they started.

The length of a linear fortification does not fit any parameter, it can be short as the 2.5km (1.6 miles) of the Hilsea Lines raised by the British to defend island of Portsea against French attack in the 1750's, to the Great wall of China 6350 kilometres (3945 miles) with its various branches dating between 220 BCE - 1368-1644 CE. To contrast the Great wall of China there is the French *Barrière de fer* (barrier of iron) which was the eastern border of France with Prussia (1874 – 1914 CE) that was 2044 km (1270 miles),

You could think that over time the sophistication of linear fortifications would evolve, going from ditch and ramparts to stone walls defended by towers, back to the earth and wood trenches of WWI, a couple decades later there are the re-enforced concrete bunkers and casemates of the Atlantic wall, and finally we regress again to trenches of the Russian / Ukrainian war.

The multiple '*stop lines*' of the Germans, a form of '*defence in depth*' did not stop the overwhelming juggernaut of the Allies through the Italian peninsula, let alone on mainland Europe. Once the allies controlled the air and provided large amounts of men, machines and supplies no linear fortification was good enough, no matter how much re-enforced concrete was poured.

An interesting comparison is the linear fortification built by counties that felt the need to build them to feel safer. The Roman Empire raised eight border defences as the empire was expanding, including the surrounding the Mediterranean Sea, whereas Germany needed eight border, coastal and stop lines to defend against the onslaught of the Allies. Countries like France and Britain were attacked multiple times, sometimes by each other, and France's Maginot Line encouraged many other countries to emulate them. Luckily England's '*stop lines*' never got tested because Germany never invaded. As the Byzantine Empire grew into more significance with the decline of Rome, and therefore needed to protect its capital Constantinople with sea walls and land walls, including the Theodosian Land walls.

On the other hand, threats by Russia caused six linear fortifications to be built in the period from the 15th century CE to the present. And Germany wins the medal tally at nine fortifications raised from the 19th century at the end of WWII.

In some cases the frontiers are still in place and they are called Demilitarized zones (DMZ or DZ). A DMZ is a designated area by agreements or treaties between countries or opposing groups that bar the building of military installations, activities of a military nature or any military personal in that area. A DMZ may be an established frontier between two countries such as North and South Korea (1953 CE), or between Iraq and Kuwait (1991 CE).

Fortifications no longer necessary are left to rot or are dismantled, some are preserved to remind people of their purpose. All fortifications require regular maintenance to stop the egress of water, to creep of nature, be it the infestation of plant life or the erosion of the sand supporting the structure. Ploughing and the march of the seasons will remove the remnants of the earthen ditches and ramparts.

The nature of warfare has changed over time from the siege engines of the Assyrian (800 BCE), the cannons used by the Turks pounding the walls of Constantinople (1453 CE), the bomber aircraft of the Allies, let alone the use of H-Bombs in 1945 CE which saw the surrender of Japan. Now we see un-manned drones blowing up aircraft and oil refineries, as well as being flown into enemy dug outs on the trenches of the Russian / Ukraine war (currently in progress as I type). All things change and nothing ever stays the same.

Bibliography.

Allen Brown, R. (1954). *English Medieval Castles.* London, United Kingdom: Batsford.

Allen Brown, R. (1985) *Castles - A History and Guide.* Dorset, United Kingdom: Blandford.

Anderson, W. (1984). *Castles of Europe.* Ware, Hertfordshire, United Kingdom: Omega Books Ltd.

Balent. M. (1985). *The Palliadium Book of European Castles.* Detroit: Palliadium Books.

Barrucand, M. & Bednorz, A. (1992). *Moorish Architecture in Andalusia.* Köln, Germany: Taschen.

Bottomley, F. (1979). *The Castle Explorer's Guide.* London, United Kingdom: Kaye and Ward.

Bradbury, J. (1992).*The Medieval Siege.* Martlesham, Great Britain: Boydell Press.

Brice, M.H. (1990). *Forts and Fortresses.* London, United Kingdom: Quarto Publishing.

Brice, M.H. (1984). *Stronghold. A History of Military Architecture.* London, United Kingdom: Chrysalis Books.

Crosby, A.W. (2002). Throwing Fire, Projectile Technology Through History. Cambridge, United Kingdom: Cambridge University Press.

Cunliffe, B. (1986). *Danebury, Anatomy of an Iron Age Hillfort.* London: BT Batsford Ltd.

Cunliffe, B. (1975). *Iron Age Communities in Britain.* London: BT Batsford Ltd.

De Breffny, B. (1977). *Castles of Ireland.* London, United Kingdom: Thames & Hudson Ltd.

Donnelly, M.P. & Diehl, D. (1998). *Siege, Castle at War.* Texas: Taylor Publishing Company.

Duby, G. (1991). *France in the Middle Ages 987 - 1460.* Malden. Massachusetts, United States of America: Blackwell Publishers.

Edwards, R.W. (1987). *The Fortifications of Armenian Cilicia* Washington D.C., United States of America: Dumbarton Oaks Research Library and Collection.

Falck, W. (2000). *A cultural and historical walk around Visby town wall.* Gotland: Gotlands Fornsal.

Fass, V. (1986). *The Forts of India.* London, United Kingdom: Collins Publishers.

Forde-Johnston, P. (1977). *Hadrian's Wall.* London, United Kingdom: J M Dent & Sons.

Foss, C and Winfield. D. (1986) *Byzantine Fortifications: An Introduction.* Praetoria, South Africa: University of South Africa.

Froissart, J. (1401-1500). *Chroniques sire Jehan Froissart.* Bibliothèque nationale de France. Département des Manuscrits. Français 2644 [was BNF 8320]

Fry, P.S. (1974). *British Medieval Castles.* London, United Kingdom: David & Charles.

Gebelin, F. (1964). *The Chateaux of France.* London, United Kingdom: Ernest Benn.

Gravett, C. (1990). *Medieval Siege Warfare.* London, United Kingdom: Osprey Publishing.

Guilbert, G., Ed. (1981). *Hill-fort studies: essays for A H A Hogg.* Leicester, United Leicester University.

Hindley, G. (1968)., *Castles of Europe (Great Buildings of the World).* Norwich, United Kingdom: Paul Hamlyn.

Hinago, M. (1986). *Japanese Castles.* Tokyo, Japan: Kodansha.

Higham, R. & Barker P. (1992). *Timber Castles.* London, United Kingdom: B.T. Batsford.
Hogg, I. (1981). *The History of Fortification.* London, United Kingdom: Orbis Publishing.
Hughes, Q. (1974). *Military Architecture.* London, United Kingdom: Evelyn.
Hughes, Q., (1991). *Military Architecture, The Art of Defence from Earliest Times to the Atlantic Wall.* Beaufort Publishing.
Johnson, A. (1983). *Roman Forts of the 1st and 2nd centuries AD in Britain and German Provinces.* London, United Kingdom: Adam & Charles Black.
Johnston, J.F. (1977). *Hadrian's Wall.* London: Book Club Associates.
Johnston, P. (1979). *The National Trust Book of British Castles.* London, United Kingdom: Weidenfeld & Nicolson.
Johnston, P. (1989). *Castles of England, Scotland and Wales.* London, United Kingdom: Weidenfeld & Nicolson..
Keeley, L.H., Fontana, M., & Quick, R. (2007). Baffles and Bastions: The Universal Features of Fortifications. *Journal of Archaeological Research, 15,* 55-95. DOI: http://dx.doi.org/10.1007/s10814-006-9009-0
Kennedy, H. (1995). *Crusader Castles.* Cambridge, United Kingdom: Cambridge University Press.
Kightly, C. (1979). *Strongholds of the Realm.* London, United Kingdom: Thames & Hudson Ltd.
Kontogiannis, N. (2022). *Byzantine Fortification – Protecting the Roman Empire in the East.* Barnsley, Yorkshire, United Kingdom; Havertown, Pennsylvania, United States of America: Pen and Sword Military.
Lawrence, T.E. (1986). *Crusader Castles.* London, United Kingdom: Michael Haag.
Lepage, J. (2023). *Dictionary of Fortification.* Barnsley, United Kingdom: Pen and Sword Military
Manley J. (1989). *Atlas of Prehistoric Britain.* London, United Kingdom: Phaidon.
Minney, R.J. (1971). *The Tower of London.* London, United Kingdom: Cassell.
Mishra, R, (2022). *Military architecture in ancient India.* B.R. Publishing Corporation.
Muller, J. (1968). *Treatise of Fortifications,* 1746. Ottawa, Ontario, Canada: Museum Restoration Service.
Nicolle, D. (1998). *Medieval Warfare Source Book: Christian Europe and its Neighbours.* London, United Kingdom: Brockhampton Press.
Noonan, J. (2000). *The Daily Telegraph Castles and Ancient Monuments of Scotland.* London: Aurum Press Limited.
Oman, C., Sir. (1978). *Castles,* An illustrated guide through 80 castles of England and Wales. New York, United States of America: Crown Publishers.
O'Neil, B.H. (1954). *Castles. An introduction to the castles of England and Wales.* London: Her Majesty's Stationery Office.
Rolf Sjöblom, Eva Hjärthner-Holdar, Carolyn I. Pearce, Jamie L. Weaver, Erik Ogenhall, John S. McCloy, José Marcial, Edward P. Vicenzi, Michael J. Schweiger, Albert A. Kruger, Assessment of the reason for the vitrification of a wall at a hillfort. The example of Broborg in Sweden, *Journal of Archaeological Science: Reports, Volume 43,* 2022, 103459, ISSN 2352-409X, https://doi.org/10.1016/j.jasrep.2022.103459
Ross, S. (1990). *Scottish Castles.* Moffat, Scotland: Lochar Publishing.
Rybot, N. (1960). *The Islet of St. Helier and Elizabeth Castle – Jersey.* Jersey: States of Jersey.

Saunders, A. (1989). *Fortress Britain, Artillery fortification in the British Isles and Ireland.* Liphook, United Kingdom : Beaufort.
Sorrel, A. (1973). *British Castles.* London, United Kingdom: Hastings House
Smith, Capt. G. (1969). *An Universal Military Dictionary,* 1779. Ottawa, Ontario, Canada: Museum Restoration Service.
Stierlin, H. (2006). *Unfolding History – Castles, Fortresses and Citadels.* London: Thames and Hudson.
Sullivan, D.F. (2000). *Siegecraft, Two Tenth Century Instruction Manuals by "Heron of Byzantium".* Washington D.C., United States of America: Dumbarton Oaks Research Library.
Tabarelli, G.M. (1979). "Ideal" and Fortified Cities of the Renaissance. In Held R. [editor] (1979). *Arts, Arms and Armour, An International Anthology Volume 1, 1979 - 80.* Switzerland, Aquafresca editrice SA.
Thompson, H., A., (1912). *Military Architecture in England during the Middle Ages.* Oxford University Press.
Toy, S. (1955). *A History of Fortifications from 3000 BC to AD 1700.* London, United Kingdom: Heinemann.
Tsangadas, B.C.P. (1980). *The Fortifications and Defence of Constantinople.* East European monographs, No. 71, New York, United States of America: Columbia University Press.
Tulse, A. (1958). *Castles of the Western World..* London, United Kingdom: Thames and Hudson.
Viollet-le-Duc, E.E. (1990). *Military Architecture.* Greenhill Books.
Warner, P. (1968). *Sieges of the Middle Ages.* London, United Kingdom: G. Bell and Sons.
Webster's Revised Unabridged Dictionary (1996, 1998). New York, New York, United States of America: Random House Reference; MICRA, Inc.
Weissmüller, A.A. (1967). *Castles From the Heart of Spain.* London, United Kingdom: C. N. Potter.
Wyley, S.F. (1996). *Dictionary of Military Architecture.*
Wyley, S.F. (1998). *What is a Castle?* In *Varangian Voice, Issue 50, February 1999.* Melbourne, Victoria, Australia: New Varangian Guard.

Further Reading.

Campbell D, B. (2002). *Aspects of Roman Siege Craft.* [Doctoral thesis, Department of Archaeology, University of Glasgow]. glathesis:2002-776 https://theses.gla.ac.uk/776/

Constantinople and its Hinderland: Papers from the Twenty-seventh Spring Symposium of Byzantine studies, Oxford, April 1993. #9 "The Long Walls of Thrace" by J.G Crow (pages 109-124)

A.M. Gubaidullin. *Fortification Dictionary.* 2nd edition, updated and revised. Kazan: Institute of History of the Tatarstan Academy of Sciences, 2006. http://archtat.ru/en/editions/a-m-gubaidullin-fortification-dictionary-2nd-edition-updated-and-revised-kazan-institute-of-history-of-the-tatarstan-academy-of-sciences-2006-144-p/

Fox, A. (1976). *Prehistoric Maori fortifications in the North Island of New Zealand.* Auckland, New Zealand: Longman Paul

Hallé, Guy le (1986). [*Dictionary and Terminology of Fortification] Précis de la Fortification*. Lyon, France: Editions Horvath.

Kaufann, J.E. (2024). *Forts and Fortifications of Europe, 1815-1945:The Central States: Germany, Austria-Hungary and Czechoslovakia*. Barnsley, United Kingdom: Pen and Sword Military.

Kaufann, J.E. and Kaufann, H.W. (2022). *Fortress Europe: From Stone to Steel Fortifications,1850–1945.* Barnsley, United Kingdom: Pen and Sword Military.

Kaufann, J.E. and Kaufmann, H.W. (2019). *Castle to Fortress: Medieval to Postmodern Fortifications in the Lands of the Former Roman Empire.* Barnsley, United Kingdom: Pen and Sword Military.

Kaufann, J.E. and Kaufmann, H.W. (2018). *Castrum to Castle: Castrum to Castle: Classical to Medieval Fortifications in the Lands of the Western Roman Empire.* Barnsley, United Kingdom: Pen and Sword Military.

Kaufann, H.W. and Kaufann, J.E. (2016). *Verdun 1916: The Renaissance of the Fortress.* Barnsley, United Kingdom: Pen and Sword Military.

Kaufann, J.E. ,Kaufann, H.W. Potocnik, A. and Lange, P. (2011). *The Maginot Line: History and Guide.* Havertown, Pennsylvania, United States: Casemate Publishers.

La Cour, V. & Stierdahl, H. (Volume 1 1956, Volume 2, 1963). *Danske Voldstedor fra Oldtid og Middelalder.*

Lawrence A. W. (1983). A skeletal history of Byzantine fortification. *Annual of the British School at Athens, 78.* pp. 171–227..

Maurice (582-602 AD). *Strategicon.* Maurikos; George T. Dennis, translator. (1984). *Maurice's Strategikon. Handbook of Byzantine Military Strategy, translated from Greek to English by George T. Dennis.* Philadelphia, Pennsylvania, United States of America: University of Pennsylvania Press. At Internet Archive, https://archive.org/details/maurice-strategikon-dennis-1984/Maurice_Strategikon_Dennis_1984/

Nikos D. Kontogiannis. (2022). *Byzantine Fortifications. Protecting the Roman Empire in the East.* Barnsley, United Kingdom: Pen & Sword Military.

Orai, P, & Toba M. (1935). *Castles in Japan / by N. Orui and M. Toba.* Tokyo, Japan: Board of Tourist Industry, Japanese Government Railways.

Renn, D. F. (1959). Mottes: A Classification. *Antiquity, 33*(130), 106–112 . Cambridge, United Kingdom: Cambridge University Press. DOI: https://doi.org/10.1017/S0003598X00029136

Rolf, R. (2004). *A Dictionary on Modern Fortification. Fortification - An illustrated lexicon on European Fortification in the period 1800-1945.* Middelburg, Holland: PRAK Publishing.

Rottman, Gordon L., Palmer Professor Ian [Illustrator] (2012, May 20). *Osprey Fortress Series: 23 German Field Fortifications 1939-45.* Oxford, United Kingdom: Osprey Publishing.

Rottman, Gordon L., Palmer Professor Ian [Illustrator] (2012, May 20). *Osprey Fortress Series 62: Soviet Field Fortifications 1941-45.* Oxford, United Kingdom: Osprey Publishing.

Links.

American civil war.

Chuber, D. C. (1996). *Field fortifications during the American civil war: a tactical problem.* [Master's thesis, Fort Leavenworth, Kansas]. Defense Technical Information Centre. https://apps.dtic.mil/sti/pdfs/ADA313032.pdf

Tennessee Secretary of State Tre Hargett. (2018). *Civil War Source Book - Glossary.* Nashville, Tennessee, United States of America : Tennessee State Library and Archives.
https://www.tnsos.net/TSLA/cwsourcebook/glossary.php

UWRF Archives. (n.d.). Glossary. *The Civil War and Northwest Wisconsin.*
https://thecivilwarandnorthwestwisconsin.wordpress.com/glossary/

Definitions, Dictionaries, Glossaries.

Thomas, J. L. (2009). *Definition of a Castles compiled for the Castles of Wales by Jerome Morris: What is a Castle?* The Castles of Wales. Retrieved April 29, 2024 from https://www.castlewales.com/cast_def.html.

Spiteri, S. C. (2010). Illustrated glossary of terms used in military architecture terms. *Arx: International Journal of Military Architecture and Fortification,* 635-650. L-Università ta' Malta.
https://www.um.edu.mt/library/oar/handle/123456789/90003

Historic England (2017, December). *Military Structures: Listing Selection Guides.* https://historicengland.org.uk/images-books/publications/dlsg-military/heag123-military-structures-lsg/.

Weaver, John R. II. (2021, June 7). Glossary of Fortification Terms as used in the American Third System of Masonry Coastal Forts, 1816-1867. CivilWarTalk [forum]. Retrieved 2 July, 2024 from
https://civilwartalk.com/threads/glossary-of-fortification-terms-as-used-in-the-american-third-system-of-masonry-coastal-forts-1816-1867.185896/

New York State Military Museum and Veterans Research Center. (n.d.). *Glossary of terms.* Retrieved 2 July, 2024 from
https://museum.dmna.ny.gov/application/files/3615/9846/3622/Glossary_Forts.pdf

Fort Adams and the Fort Adams Trust. (n.d.). *The terminology of a fortress.* Fort Adams. https://fortadams.org/discover-the-fortress/the-terminology-of-a-fortress/

The American Battlefield Trust (2018, March 28, updated 2023, October 17). *A glossary of fortification terms: A list of fortification terminology.* https://www.battlefields.org/learn/articles/glossary-fortification-terms

Hypogean Archaeology. (n.d.). *Typology 6: military structures.* Retrieved 12 July, 2024 from https://www.archeologiadelsottosuolo.com/wp-content/uploads/2020/11/Typology-6.pdf

Camouflage.

Elias, Dr Ann. (2003, April). The organisation of camouflage in Australia in the Second World War. *Journal of the Australian War Memorial, 38.* Canberra, ACT, Australia : Australian War Memorial.
https://www.awm.gov.au/articles/journal/j38/camouflage

Fortification, Works and Castles.

Tourist Jordan. (n.d.) *Ajloun castle.* https://www.touristjordan.com/ajloun-castle/

Wikipedia contributors. (2025, July 30). British hardened field defences of World War II. In *Wikipedia, The Free Encyclopedia.* Retrieved 04:28, August 17, 2025, from https://en.wikipedia.org/w/index.php?title=British_hardened_field_defences_of_World_War_II&oldid=1303348011

Admin. (2025, July 5) Bunkers, Dykes, and a Dash of Panic: A Jolly Jaunt Through Dutch Fortifications in the Five-Day Fisticuffs of 1940! *Bayonets & Brushes* https://bayonetsandbrushes.co.uk/bunkers-dykes-and-a-dash-of-panic-a-jolly-jaunt-through-dutch-fortifications-in-the-five-day-fisticuffs-of-1940/

Deloche, J. (2007). III. The Muslim System of Fortification in the Deccan Kingdoms (15th-18th century). In Studies on fortification in India (1–). Institut Français de Pondichéry. https://doi.org/10.4000/books.ifp.4020

Wessex Archaeology. (2006, January)
Concrete Sound Mirror, Cinque Ports Training Area Hythe, Kent
Final Report
https://www.wessexarch.co.uk/our-work/concrete-sound-mirror-cinque-ports-training-area

FORTIFICATION AND SIEGECRAFT EB 1911 Editor A.W.J Graham Kerr & picture editor C.H blackwood
RESEARCH guide Vi1. 2020 Encyclopaedia BRITANNICA 1911
https://ordnancesociety.org.uk/wp-content/uploads/2020/09/os-research-guide-vii-fortification-and-siegecraft.pdf

Lawrence, A. W. (1965). *Ancient Egyptian Fortifications.* The Journal of Egyptian Archaeology, Vol. 51 (Dec., 1965), pp. 69-94 (28 pages). Published By: Sage Publications, Ltd.
https://www.jstor.org/stable/3855621

Mizokami, K. (2020, December 31). Medieval Star Forts Are Surprisingly Alive and Well in North Africa - We haven't seen these in a couple centuries. *Popular Mechanics.* New Youk, United States: Hearst Digital Media.

https://www.popularmechanics.com/military/weapons/a35084150/vauban-fortification-star-forts-french-army-north-africa/

Current Archaeology. (2023, September 2). *Another Antonine Wall fortlet found*
https://the-past.com/news/another-antonine-wall-fortlet-found/

Pak Weather. (2015, December 4). *Ranikot Fort - The Great Wall of Pakistan.* Internet Archive
https://web.archive.org/web/20150414191808/http://www.pakweather.com/2015/04/ranikot-fort-great-wall-of-pakistan.html
UK Second World War Heritage group. (2021, August 10) Anti tank blocks. *UK Second World War Heritage.* https://ukswwh.wordpress.com/2021/08/10/anti-tank-blocks/

Chinese.
春秋戰國. (2025, June 9). Chinese fortification: an overview of parts and terminology — Part 1: The wall. *Great Ming Military*
https://greatmingmilitary.blogspot.com/2019/06/chinese-fortification-p1.html

春秋戰國. (2019, June 13). Chinese fortification: an overview of parts and terminology — Part 1: The wall. *Great Ming Military.*
https://greatmingmilitary.blogspot.com/2019/06/chinese-fortification-p1.htmll
春秋戰國. (2019, June 27). Chinese fortification: an overview of parts and terminology — Part 2: Gate and moat. *Great Ming Military.*
https://greatmingmilitary.blogspot.com/2019/06/chinese-fortification-p2.html

Forts.
Lopez, Miguel A. (2017, November 17). Forts. *International Encyclopedia of the First World War.* https://encyclopedia.1914-1918-online.net/article/forts

Fougasse.
Ghajnsielem.com. (n.d.). *The Fougasse.* Retrieved 17 January, 2024 from https://www.ghajnsielem.com/places/fougasse.html

Guédelon Castle, a castle based on 13th century design which is currently under construction near Treigny, France.
https://www.guedelon.fr/en/

Guldursun-Kala.
Guldursun-Kala. (2023, October 3). In *Wikipedia.*
https://en.wikipedia.org/wiki/Guldursun-Kala

Japanese castles.
Diego. (2020, May 1). Field report: the ruins of Suncheon Japanese castle, South Korea (26 January 2020) – part 1. *Within Striking Distance* [blog]. https://ittekuru.com/2020/05/01/field-report-the-ruins-of-suncheon-japanese-castle-south-korea-26-january-2020-part-1/
Diego. (2020, May 1). Field report: the ruins of Suncheon Japanese castle, South Korea (26 January 2020) – part 2. *Within Striking Distance* [blog]. https://ittekuru.com/2020/05/01/field-report-the-ruins-of-suncheon-japanese-castle-south-korea-26-january-2020-part-2/

Diego. (2020, May 1). Field report: the ruins of Suncheon Japanese castle, South Korea (26 January 2020) – part 3. *Within Striking Distance* [blog]. https://ittekuru.com/2020/05/01/field-report-the-ruins-of-suncheon-japanese-castle-south-korea-26-january-2020-part-3/
Spiteri, Stephen C. (2003). Glossary of Terms used in Military Architecture. *Fortress Explorer*. Retrieved 27 June, 2024 from
https://www.fortressexplorer.org/glossarya.html
Jay Castle. (2021, April 20). Trenches. *Jcastle Guide to Japanese Castles*. Retrieved 26 April 2024 from https://www.jcastle.info/view/Trenches

Licence to crenalate.

Davis, P. (2006-7). English licences to crenellate 1199-1567. *Castle Studies Group, Journal 20* on Gatehouse Gazetteer.
https://www.gatehouse-gazetteer.info/LtoCren.html

Davis, P. (2009). English licences to crenellate further analysis. *Gatehouse Gazetteer*.
https://www.gatehouse-gazetteer.info/LtoCren2.html

Linear fortification.

Dutchy Media (2024). *Benin moat Edo state*. Naijna 7 Wonders.
https://www.naija7wonders.com/benin-moat-edo-state/

Denman, D. S. (2019, December 13). On Fortification: Military Architecture, Geometric Power, and Defensive Design. *Security Dialogue, 51*(2-3), 231-247. https://doi.org/10.1177/0967010619889470

Denman, D. S. (2019). On fortification: Military architecture, geometric power, and defensive design. *Security Dialogue*, *51*(2-3), 231-247. https://doi.org/10.1177/0967010619889470 (Original work published 2020).

Fortification lines [category] (n.d.) In *Wikipedia*. Retrieved August 27, 2025 from
https://en.wikipedia.org/wiki/Category:Fortification_lines

MacDonald, C.B., Guilmartin, J.F., Baumer, W.H. (2025, May 29). *fortification. Encyclopedia Britannica*. https://www.britannica.com/technology/fortification

Miechowicz, L. (2020). Medieval fortification systems in the form of linear embankments in present-day Poland. In Drnovský, P. Hejhal, P. (2020). *Archaeology of Conflicts*. Czechoslovakia, Červený Kostelec: Pavel Mervart Publishing, pp27-39.
Accessed from:
https://www.academia.edu/50764182/Medieval_fortification_systems_in_the_form_of_linear_embankments_in_present_day_Poland

Parton, Phillip, Clark, Geoffrey, and Reepmeyer, Christian (2022) *High-resolution lidar analysis of the Fisi Tea defensive earthwork at Lapaha, Kingdom of Tonga*. In: Clark, Geoffrey R., and Litster, Mirani, (eds.) Archaeological Perspectives on Conflict and

Warfare in Australia and the Pacific. Terra Australis (54). Australia, Canberra, ACT: ANU Press, pp. 147-169. https://doi.org/10.22459/TA54.2021.08

Sarantis, A. C., & Christie, N. (2013). Fortifications in the West: a bibliographic essay. In A. Sarantis, & N. Christie (Eds.). *War and Warfare in Late Antiquity: Current Perspectives* (pp. 255-296). (Late Antique Archaeology; Vol. 8.1-8.2). Netherlands, Leiden: Brill. https://doi.org/10.1163/9789004252585_008

Shao, M., Liang, X., Wang, Y. *et al.* An overview of some traditional materials used in linear fortification structures in coastal regions of China: implications for their analytical study and conservation. *Bull Eng Geol Environ* 82, 464 (2023). https://doi.org/10.1007/s10064-023-03477-w

Shao, M., Liang, X., Wang, Y., Wang, Z., & Li, L. (2023). An overview of some traditional materials used in linear fortification structures in coastal regions of China: implications for their analytical study and conservation. *Bulletin of Engineering Geology and the Environment*, *82*(12). https://doi.org/10.1007/s10064-023-03477-w

Sîrbu, V., Croitoru, C., Pandrea, S. (2013)
The Linear Fortification from Şuţeşti, Brăila County, în (2014) *Ancient Linear Fortifications on the Lower Danube, Proceedings of the National Colloquium "Ancient Linear Fortifications on the Left Bank of the Lower Danube"*, Brăila, 14-16 June 2013. Romania, Cluj-Napoca: Editura Mega., pp. 41-56 Accessed from https://www.researchgate.net/publication/387954960_The_Linear_Fortification_from_Sutesti_Braila_County_in_vol_Ancient_Linear_Fortifications_on_the_Lower_Danube_Proceedings_of_the_National_Colloquium_Ancient_Linear_Fortifications_on_the_Left_Bank_of_th

Machine gun nest.
World War II machine gun nest uncovered in Slovakia. (2024, January 2). Archaeology Magazine. https://www.archaeology.org/news/12031-240103-slovakia-machine-gun-nest

Minefield Control Tower.
Council for British Archaeology (2006) Minefield Control Tower. in https://archaeologydataservice.ac.uk/archives/view/dob/ai_full_r.cfm?refno=442 in *Defence of Britain Archive* [data-set]. York: Archaeology Data Service [distributor] https://doi.org/10.5284/1000327

Other.

Denman, D. (2019). On fortification: Military architecture, geometric power, and defensive design. *Security Dialogue*, *51*(2-3), 231-247. https://doi.org/10.1177/0967010619889470

World Heritage Centre. (n.d.). Su Nuraxi di Barumini in the World *Heritage List*. https://whc.unesco.org/en/list/833/

JAANUS is the on-line Dictionary of Japanese Architectural and Art Historical Terminology compiled by Dr. Mary Neighbour Parent. https://www.aisf.or.jp/~jaanus/

US Army | Archive of Military Doctrine. FM 5 – Engineers Series. (1072, June 27). FM 5-15: Field Fortifications. https://www.bits.de/NRANEU/others/amd-us-archive/FM5-15%2868%29.pdf [dead link 9 October 2025]

Zammit, V. (1982). *Maltese fortifications. Civilization, 1*, 22-25. United States, Houston, Texas: Gulf Publishing Ltd. https://www.um.edu.mt/library/oar/handle/123456789/41143

Camouflaged tree.

Imperial War Museum. (n.d.) *The Journey Of The Camouflage Tree.* https://www.iwm.org.uk/history/the-journey-of-the-camouflage-tree

Imperial War Museum. (n.d.) *Why the British Army built fake trees during the First World War.* https://www.iwm.org.uk/history/why-the-british-army-built-fake-trees-during-the-first-world-war

Vigels.

www.in2greece.com, (2022). *Vigles, the medieval watchtowers of Chios.* https://www.in2greece.com/english/places/summer/islands/chios/vigles-of-chios.html

Roman.

Campbell, D.B. (2002). *Aspects of Roman Siegecraft*. Department of Archaeology, University of Glasgow. https://theses.gla.ac.uk/776/1/2002campbellphhd.pdf

Jones, G. D. B. and Mattingly, D. J. (1980). Fourth-Century Manning of the 'Fossatum Africae'. *Britannia,11,* 323-326. Society for the Promotion of Roman Studies. https://doi.org/10.2307/525690

Sasanian Frontiers and Fortifications.

Simpson, St. John. (1993, February 17). *Sasanian Frontiers and Fortifications.* Talk given in the Department of Archaeology, University of Newcastle, 17th February 1993. Retrieved 1 May, 2024 from https://www.academia.edu/4694474/Sasanian_Frontiers_and_Fortifications.

Siegfried Line.

Johnson, Patrick M. (2022, November 8). The 300 mm Thick Steel Turrets on the Siegfried Line. *Tank Historia.*https://tankhistoria.com/wwii/20p7-panzerturm/

Siege castle.
Stamper, P. A., Timby, J. R. (1984). Excavations on a mid-twelfth century siege castle at Bentley, Hampshire. *Proceedings of the Hampshire Field Club and Archaeological Society, 40*, 1984, 81-89.
https://www.hantsfieldclub.org.uk/publications/hampshirestudies/digital/1980s/vol40/Stamper.pdf

Tank trap.

Shields, A. Morris MHR, Hon. P. (1996, December 20). Belmont Tank Traps. *NSW War Memorials Register*.
https://www.warmemorialsregister.nsw.gov.au/memorials/belmont-tank-traps

Trench digging machine.

Cultivator No. 6. (2025, June 13) In Wikipedia.
https://en.wikipedia.org/wiki/Cultivator_No._6

British Pathe. (1910-1919) Mechanical trench digging. [video] Footage of mechanical trench digger in action.
https://www.britishpathe.com/asset/99070/

Malyasov, J. (2016, July 7). In Ukraine developed new PZM-3 armored trencher machine. Defence Blog. https://defence-blog.com/in-ukraine-developed-new-pzm-3-armored-trencher-machine/

Trench warfare.
Kempf, P., Clelland, C [editor]. (n.d.). WW1 Trench Schematics. *Landships II*. Retrieved 12 April 2024 from
http://landships.info/landships/fortification_articles.html?load=fortification_articles/Trench_Schematics.html

Salem Media. (2000-2024). WW1 Trenches: The Heart Of Battle! How Did They Work? *History on the Net*. Retrieved. May 10, 2024 from
https://www.historyonthenet.com/ww1-trenches-what-is-a-trench

Waldron, Captain William H. (1917). *Elements of Trench Warfare*. Project Gutenberg, February 6, 2020 [EBook #61330].
https://www.gutenberg.org/files/61330/61330-h/61330-h.htm

WWII German Wehrmacht trenches, shelters, field fortifications.
https://festung.net/wwii-german-wehrmacht-trenches-shelters-field-fortifications/

Vauban.

Association Réseau des sites majeurs de Vauban. (n.d.) Network of Vauban's major sites. https://sites-vauban.org/en

Drawings Index.

Drawings	*Credit*
Agger	Steven Lowe
Alatorium	Steven Lowe
Arrow headed bastion	Steven Lowe
Arrow loop	Katie Roberts
Arrow slit	Katie Roberts
Balistraria	Steven Lowe
Barbican	Katie Roberts
Barrage balloon	Katie Roberts
Bartizan - Fort George, Inverness, Scotland	Katie Roberts
Bastion	Steven Lowe
Bastioned tower	Stephen Wyley
Bastle	Katie Roberts
Battlement	Steven Lowe
Bergfried	Katie Roberts
Berm	Steven Lowe
Blind gun port	Steven Lowe
Bottom	Steven Lowe
Box machicolation	Katie Roberts
Broch	Steven Lowe
Caponier	Steven Lowe
Caravaserai	Stephen Wyley
Casatorri	Katie Roberts
Casemate	Katie Roberts
Chain tower	Stephen Wyley
Cheng pu	Stephen Wyley
Circumvallate	Katie Roberts
Citadel	Steven Lowe
Cliff fort	Steven Lowe
Clover leaf keep	Steven Lowe
Concentric castle, Caerphilly castle, Wales, UK.	Katie Roberts
Concrete battleship, Fort Drum, Philippines	Stephen Wyley
Corner flanker	Steven Lowe
Crannog	Katie Roberts
Crenel shutter	Steven Lowe
Crownwork	Steven Lowe
Cruciform arrow slit	Katie Roberts
Cupola	Katie Roberts
Denti di drago	Stephen Wyley

Ditch profiles; Roman, Iron age hill fort, WWI trench	Katie Roberts
Dombunker	Stephen Wyley
10th century donjon, *at Langeais, France.*	Steven Lowe
12-13th century donjon, *Château de Falaise, Normandy, France.*	Steven Lowe
Dou yan	Stephen Wyley
Dragons teeth	Katie Roberts
Drawbridge	Katie Roberts
En bec	Steven Lowe
Escaraguaita	Steven Lowe
Fire control tower	Stephen Wyley
Flying bridge	Steven Lowe
Fortified bridge	Katie Roberts
Fortified harbour	Katie Roberts
Front Todt	Stephen Wyley
Gabion	Katie Roberts
Genoese tower	Stephen Wyley
Gran bugue	Steven Lowe
Great gatehouse	Steven Lowe
Great wall of China	Katie Roberts
Gulaigorod	Katie Roberts
Gun port	Katie Roberts
Han cheng	Stephen Wyley
Hangburg	Stephen Wyley
Hemmbalk	Stephen Wyley
Hersillon	Steven Lowe
Hexco barrier	Katie Roberts
Honmaru and Ninomaru - Osaka castle	Steven Lowe
Ishigaki	Steven Lowe
Ishi-otoshi mado	Steven Lowe
Italian merlon	Steven Lowe
Kôrai mon	Steven Lowe
Lilia	Katie Roberts
Listening posts	Katie Roberts
L-plan castle	Stephen Wyley
Bunkers from the Mareth line.	Stephen Wyley
Martello tower	Stephen Wyley
Masugata mon	Steven Lowe
Maximillian tower	Katie Roberts

Merlons - various	Steven Lowe
Mile castle	Steven Lowe
Motte & bailey castle	Steven Lowe
Multivalate hill fort, Maiden castle, Dorchester, UK.	Kate Roberts
Ninomaru	Steven Lowe
Nissen huts	Katie Roberts
Oilet	Steven Lowe
Pele	Katie Roberts
Pill boxes, No. 24 and 25.	Gary Bannister
Ping tou qiang, town wall of Sanhe, Anhui, China	Stephen Wyley
Portcullis	Steven Lowe
Postern gate	Steven Lowe
Quadrangular castle	Steven Lowe
Ravelin	Steven Lowe
Redan, Reden	Steven Lowe
Rice Blockhouse	Stephen Wyley
Rivierkazemat	Stephen Wyley
Salient, Salient (blunt)	Steven Lowe
Scarp	Steven Lowe
Shell keep	Steven Lowe
Siege ladder	Stephen Wyley
Stanton Air Raid shelter	Stephen Wyley
Star fort	Katie Roberts
Talus	Steven Lowe
Tenshu	Steven Lowe
Tian jing	Stephen Wyley
Tobruk	Stephen Wyley
Torenfort	Stephen Wyley
Torre de homenaje	Steven Lowe
Tower house – 'L' plan	Steven Lowe
Umadashi	Stephen Wyley
Vepshi	Stephen Wyley
Wagonberg	Katie Roberts
Wall walk	Steven Lowe
Watergate	Stephen Wyley
Wicket	Steven Lowe
Winkle bunker	Stephen Wyley
Yagura a top an ishigaki	Steven Lowe
Z-plan tower house	Katie Roberts
Zwinger	Katie Roberts

Photographs Index.

Photos	*Credit*
Acropolos, Southern slope, Athens, Greece.	Alex de Vos
Arrow embrasure (pointed), Guldursun-Kala in modern Karakalpakstan, Uzbekistan.	Damien Fegan
Artillery tower, dating from 1636 CE, Menton, France.	Jonathon Lyons
Barbican. Caen Castle, France.	Alan Greenway
Bartizan (inside view). Fort George, Inverness, Scotland.	Wayne Robinson
Bartizan (external view). Fort George, Inverness, Scotland.	Wayne Robinson
Bascule bridge. Kalmar Castle in the province of Småland in Sweden.	Jason Metcalfe
Bastion spur, Fortezza Medicea in Siena built between 1561 and 1563 CE.	Jonathon Lyons
Battery tower. Vadstena Castle (Slott), Sweden, outside and inside.	Sarah Brown
Blind gun port. Two blind gun ports on the square wall tower, Topakapi palace, South of main gate, Istanbul, Turkey.	Stephen Wyley
https://commons.wikimedia.org/wiki/File:Anglo-Boer_War_Block_House.jpg Description English: Anglo-Boer War blockhouse, Wellington District. South Africa. This media shows a South African Protected Site with SAHRA file reference 9/2/106/0003. Date: 22 September 2015.	Source.Own work Author. Skapie777
Box machicolation. above closed up gateway in the walls of St Catherine's Monastery, Sinai Peninsula of Egypt.	Dr. Timothy Dawson
Brattice on a tower of Casscasone, department of Aude, region of Occitania, France.	Sarah Brown
Bridge head. Kalmar Castle in the province of Småland in Sweden.	Jason Metcalfe
Chateau Gaillard, Normandy, France.	Sarah Brown
Cistern. internal cistern in the Crusader castle at Shobak, Jordan, with rainwater inflow channel at right.	Dr. Timothy Dawson
Coffer, Agra fortress, Rajasthan, India.	Ruth Kunde
Cordon. Auxonne, France.	Alan Greenway
Corner buttress turret, Tower of London, England.	Awbrie Jean Culleton

Corner flanker, with arrow loops and gunports, Legge Mount, in the north-western corner of the outer ward. Tower of London, England	Awbrie Culleton
Courtyard, keepless or quadrangular castle, Bolton Castle, Yorkshire, England, UK, late 14th century (1399) CE	Awbrie Culleton
Crusader castle. "Pharoah's" or "Coral" Island, Gulf of Aqaba, Egypt.	Dr. Timothy Dawson
Curvilinear wall, Chateau Gaillard, Normandy, France.	Sarah Brown
Dame. Auxonne, France.	Alan Greenway
Defence in depth – Casscasonne, department of Aude, region of Occitania, France.	Sarah Brown
Defence (Lines of), The repaired (UNESCO grant) land walls of Constantinople, Istanbul, Turkey	Stephen Wyley
Diatenshu, Himeji castle, Japan.	Ruth Kunde
Ditch. Maiden castle, Dry ditches and ramparts.	Wayne Robinson
Dobei. Himeji Castle, Japan –.	Ruth Kunde
Drawbridge, Fort George, Inverness, Scotland.	Wayne Robinson
Drum towers of the Château d'Angers, city of Angers in the Loire Valley, in the département of Maine-et-Loire, in France.	Sarah Brown
'*D*' shaped flanker, dating to the Roman period, Carcassonne, France.	Jonathon Lyons
'*D*' shaped flanker. 13th century CE, Carcassonne, France.	Jonathon Lyons
Dun Telve, Glenelg, Scotland, from the outside.	Wayne Robinson
Dun Telve, Glenelg, Scotland. From the inside.	Wayne Robinson
Embrasure, Calshot Castle, Calshot Spit, Hampshire, embrasure.	Wayne Robinson
Embrasures (inside). Ajloun castle, Jordan.	Georgie Stickles
Embrasure. Marble tower, end of the Land walls of Constantinople, Embrasure in the interior and the exterior side of the embrasure in the '*Marble tower*'.	Stephen Wyley
Embrasure. Caen Castle, France.	Alan Greenway
Enceinte. Conwy wall, exterior of Mill Gate, in the southern wall.	Graeme & Kim Walker
Enceinte. Conwy wall - The south wall looking west, towers 21 to 18, including the Mill Gate.	Graeme & Kim Walker
Forebuilding. Rochester Castle, Kent, stairs leading up to the entry to the forebuilding.	Wayne Robinson
Gate. A gateway with gates covered in downwards overlapping metal plates 4-6mm thick, riveted to the gate, Soguk Çesme Sk., south west of the main gate of the Topkapi palace, Istanbul.	Stephen Wyley
Gate house, Harlech Castle, Gwynedd, Wales.	Wayne Robinson
Gate towers, Casscasone, department of Aude, region of Occitania, France.	Sarah Brown

Gateways, Beaumaris Castle, Anglesey, Wales.	Wayne Robinson
Giri-durga, Mehrangarh Fort, Jodhpur, Rajasthan, India.	Naomi Haddrick
Grilles. Nantes château, Nantes, Loire-Atlantique département, France.	Wendy Lale
Gun port with slit. Caen Castle, France, external and internal views.	Alan Greenway
Hako hazama, Himeji Castle, Japan.	Hannay May
Hirayamajiro with dobei in the foreground, Himeji Castle, Japan.	Ruth Kunde
Hoard, Caerphilly Castle, South Wales. outside the hoard.	Wayne Robinson
Hoard, Caerphilly Castle, South Wales. Inside the hoard.	Wayne Robinson
Indian merlons. Agra fortress, Rajasthan, India.	Ruth Kunde
Ishi-otoshi mado, Himeji Castle.	Ruth Kunde
Keep, Rochester castle, Kent, England.	Wayne Robinson
Keep, Conisbrough Castle late late 12th century (1180 – 1190 CE).	Awbrie Culleton
Kitchen, Krak des Chevaliers, Syria.	Georgie Stickles
Korai mon, Himeji Castle, Japan. Ruth Kunde.	Ruth Kunde
Land Walls of Constantinople. Looking north from the Yedicule at the land walls of Constantinople, Istanbul, Turkey.	Stephen Wyley
Machicolations above a gateway, and draw bridge, Caen Castle, France.	Alan Greenway
Machicolations atop of a tower, Palace of the Grand Master, Hospitaller, Rhodes, Greece.	Kristen Pincott
Merlon arrow slit in wall fronting the moat in front of the Golden Gate. Istanbul, Turkey.	Stephen Wyley
Moat, Beaumaris Castle, Anglesey, Wales.	Wayne Robinson
Moorish merlons, Alhambra, Granada, Spain.	Sarah Brown
Motte with a stone '*quarterfoil keep*' on top, Clifford's Tower, York, United Kingdom.	Sarah Brown
Mural tower on the North side of the Yedicule tower complex.	Stephen Wyley
Murder hole at the top of the gateway, Trim Castle, County Meath, Ireland.	Sarah Brown
Mushamada, Himeji Castle, Japan.	Hanah Ferrar
Nurhag, Outside the front of the megalithic fortress, Su Nuraxi di Barumini, Sardinia.	Arron Southwell
Nurhag, from the lowest access point of the tower, Su Nuraxi di Barumini, Sardinia.	Arron Southwell
Nurhag, Outside from the lowest access point of the tower, Su Nuraxi di Barumini, Sardinia.	Arron Southwell
Observation tower, Fort Nepean, Victoria, Australia.	Katie Roberts
Ogival bastion and points. Narbonne Gate, 13th century CE, East entrance to the citadel of Carcassonne, France.	Jonathon Lyons

Polygonal tower with a postern gate, between Yedicule and the Marble Tower, Istanbul, Turkey.	Stephen Wyley
Postern. At the base of the walls of "*King David's tower*", Jerusalem, Israel.	Dr. Timothy Dawson
Portcullis, Chateau Gisor, Normandy, France.	Sarah Brown
Pointed merlons. North west tower of the Yedicule, Istanbul, Turkey.	Stephen Wyley
Put log holes inside the postern gate in the corner of the Polygonal tower between the Yedicule and the Marble Tower.	Stephen Wyley
The ramparts of Fort George, Inverness, Scotland, Rampart.	Wayne Robinson
Redoubt, Marie Therese, France.	Michael Brown
Sama at Matsumoto Castle, Japan.	Rhonda Metcalfe
Sekizou no mon at Sogen ji Temple, Okinawa, Japan.	Sean McBride
Shot deflecting battlements Caerphilly Castle, South Wales, UK.	Wayne Robinson
Slight. Corfe Castle, Dorset. Slighted using explosives by order of Parliament following the English Civil Wars.	Wayne Robinson
Spur, Chepstow Castle, Monmouthshire, Wales, UK.	Wayne Robinson
Square flanking tower, Rochester Castle, Kent, UK.	Wayne Robinson
Square wall tower, Square wall tower, Sea walls under repair 1992, Istanbul, Turkey.	Stephen Wyley
Talus. Crac des Chevaliers, Syria.	Dr. Timothy Dawson
Theodosian (Land) Walls, Istanbul, Turkey	Stephen Wyley
Tower House. Dysert O'Dea Castle, Mollaneen, County Clare, Ireland.	Steve Sinclair
Trebuchet. '*Counterweight*' trebuchet, Caerphilly castle.	Wayne Robinson
Tunnel. Krak des Chevaliers, Syria.	Georgie Stickles
Tunnels running through a rampart. Auxonne, France.	Alan Greenway
Turreted, Rochester Castle, Kent.	Wayne Robinson
Uchikomihagi, Ishigaki at Himeji Castle.	Adrian Ferrar
Ward, Beaumaris Castle, Wales.	Sarah Brown
Watari yagura, . Hishi gate, Himeji Castle, Japan.	Ruth Kunde.
Wicket, in the main gate of the Grand Master's Palace, Valletta, Malta.	Aaron Southwell
Yagura, Tokyo Imperial palace.	Adrian Ferrar
Yagura mon, Himeji Castle, Japan.	Ruth Kunde.
Yett, Doune Castle, Perthshire, Scotland.	Wayne Robinson
Zharakty gate, Otrar, Kazakhstan.	Damien Fegan
Wall tower facing the Sea of Marmara being repaired – Spot the Author.	A. Nonimouse.

About the author – Stephen '*Sven*' Wyley.

Wall tower facing the Sea of Marmara being repaired (1992).

Stephen Wyley was born in the town of Colac ,Victoria, Australia and now lives in Melbourne (Victoria). The eldest son of a nurse and a fibrous plasterer, worked for twenty five years in analytical and research science for organisations such as the EPA and CSIRO. Later working in occupational health and safety because safety paid more and he had a desire to make the world a safer place one workplace at a time. After another twenty years in safety (thirteen years as a state government safety inspector) is now a self-employed author and craftsperson.

A living historian and re-enactor since joining the New Varangian Guard Inc. in 1984 CE. Founding '*Sven the Merchant*', a business making and selling medieval goods mainly consisting of furniture, arrows and leatherwork in 2004[1] CE.

And now is passing on this legacy in the form of in person workshops and a series of '*How-to*" historical replica construction books, along with other titles.

[1] Stephen Francis Wyley, not to be confused with his cousin or an actor in Ireland of the same name. https://www.imdb.com/name/nm3184249/
Further information can be found at:
https://sites.google.com/site/svenskildbiter/
https://www.youtube.com/channel/UCYaAcfT70zRoQv2RzNC829A
https://www.patreon.com/SventheMerchant

Forthcoming volumes.

Historical replica construction:
Viking, Volume 3;
Byzantine, Volume 1;
16th century, Volume 1;
Roman, Volume 1;
Anglo-Saxon, Volume 1.

Other titles in the works.

A History of Tents - Ancient to the 15th century.

The Big book of Boxes and Chests (for all those that did not make it into the other books).

A book of short stories of horror, science fiction, Varangians, sport, and a cat.

Previous volumes.

Historical replica constructions in Wood and Metal, Viking Volume 1 by Wyley, Robinson and Joyce, 2nd Issue, 2023, Melbourne, Intertype.

https://booksonlineaustralia.com.au/shop/historic-viking-replicas-vol-1/

Historical replica constructions, 14th Century Volume 1 by Wyley, Fraser and Robinson, 2023, Melbourne, Intertype.

https://booksonlineaustralia.com.au/shop/historic-replica-construction/

Historical replica constructions, Viking Volume 2 by Wyley, Boyle, Delany, Henry, Joyce, Robinson, and Sinclair 2024, Melbourne, Intertype.

https://booksonlineaustralia.com.au/shop/historic-replica-constructions-vikings-volume-2/

Dictionary of Military Architecture, Fortifications, Field works and Siegecraft by Wyley 2024, Melbourne, Intertype.

https://booksonlineaustralia.com.au/shop/dictionary-of-military-architecture/

Historical replica constructions, 13th Century Volume 1 by Wyley, Baker, Boyle, Erakovic, Henry, Manzi, 2025, Melbourne, Intertype.

https://booksonlineaustralia.com.au/shop/historic-replica-construction-13th-century-volume-1/

Made in the USA
Monee, IL
05 November 2026

67997471R10234